D0322209

A NEW SERIES

Synopses of the British Fauna

Edited by Doris M. Kermack and R. S. K. Barnes

e *Synopses of the British Fauna* are illustrated field and laboratory pocket-books designed to eet the needs of amateur and professional naturalists from sixth-form level upwards. Each lume presents a more detailed account of a group of animals than is found in most field-guides d bridges the gap between the popular guide and more specialist monographs and treatises. echnical terms are kept to a minimum and the books are therefore intelligible to readers with no evious knowledge of the group concerned.

olumes 1–18 inclusive are available only from the Linnean Society, Burlington House, Piccadilly, ondon W1V 0LQ

1. *British Ascidians* R. H. Millar
2. *British Prosobranchs* Alastair Graham
3. *British Marine Isopods* E. Naylor
4. *British Harvestmen* J. H. P. Sankey and T. H. Savory
5. *British Sea Spiders* P. E. King
6. *British Land Snails* R. A. D. Cameron and Margaret Redfern
7. *British Cumaceans* N. S. Jones
8. *British Ophisthobranch Molluscs* T. E. Thompson and Gregory H. Brown
9. *British Tardigrades* C. I. Morgan and P. E. King
10. *British Anascan Bryozoans* J. S. Ryland and P. J. Hayward
11. *British Freshwater Bivalve Mollusca* A. E. Ellis
12. *British Sipunculans* P. E. Gibbs
13. *British and Other Phoronids* C. C. Emig
14. *British and Ascophoran Bryozoans* P. J. Hayward and J. S. Ryland
15. *British Coastal Shrimps and Prawns* G. Smaldon
16. *British Nearshore Foraminiferids* John W. Murray
17. *British Brachiopods* C. H. C. Brunton and G. B. Curry
18. *British Anthozoans* R. L. Manuel
19. *British Planarians* I. R. Ball and T. B. Reynoldson
20. *British Pelagic Tunicates* J. H. Fraser
21. *British and Other Marine and Estuarine Oligochaetes* R. O. Brinkhurst
22. *British and Other Freshwater Ciliated Protozoa: Part I* C. R. Curds
23. *British and Other Freshwater Ciliated Protozoa: Part II* C. R. Curds, M. A. Gates and D. McL. Roberts
24. *British Nemerteans* R. Gibson
25. *Shallow-water Crabs* R. W. Ingle
26. *Polyclad Turbellarians* S. Prudhoe

TANAIDS

No. 27

TANAIDS

Keys and notes for the identification of the species

D. M. HOLDICH and J. A. JONES

Department of Zoology, The University, Nottingham NG7 2RD

1983

Published for

The Linnean Society of London

and

The Estuarine and Brackish-water Sciences Association

by

Cambridge University Press

Cambridge

London New York New Rochelle

Melbourne Sydney

Published by the Press Syndicate of the University of Cambridge
The Pitt Building, Trumpington Street, Cambridge CB2 1RP
32 East 57th Street, New York, NY 10022, USA
296 Beaconsfield Parade, Middle Park, Melbourne 3206, Australia

First published 1983

Printed in Great Britain at The Pitman Press, Bath

Library of Congress catalogue card number: 82–12761

British Library Cataloguing in Publication Data

Holdich, D. M.
Tanaids. – (Synopsis of the British fauna. New series; no. 27)
1. Tanaidacea
I. Title II. Jones, J. A. III. Linnean
Society of London IV. Estuarine and
Brackish-water Sciences Association
595.3′74 QL444.M38

ISBN 0 521 27203 3

A Synopsis of the Tanaids

D.M. HOLDICH and J.A. JONES
Department of Zoology, The University, Nottingham NG7 2RD

Contents

Foreword

Tanaid crustaceans are poorly known, especially when compared with their peracaridan relatives, the mysids, cumaceans, isopods and amphipods. Most appear in collections when they are least expected, yet they occur with great regularity in samples taken from a wide variety of habitats: they live amongst coralline algae in rock-pools, in mud, in the crevices between the plates on the back of turtles, in the abyssal trenches and some are found in fresh water. At depths of 5000 m they comprise one-fifth of the macrobenthic biomass.

When described in text-books, tanaids often appear as 'chimaeras' with the superficial features associated with decapods, isopods, amphipods and others. In other words they have an ungainly appearance when dead or in the preserved state but this image is quickly dispelled when they are seen alive. Unfortunately, the privilege of so doing has not been granted to many naturalists as tanaids are small retiring animals which are easily overlooked.

The authors of this *Synopsis* have by their clear illustrations and interesting accompanying text given us a true picture of the tanaids. Let us hope that these fascinating small crustaceans will quickly lose the bizarre image, which has so unkindly been given them in the past.

The *Synopsis of the British Fauna* are practical field-guides for those who 'go into the field' and see animals as they really are. Their now familiar format, with waterproofed covers and spaces for field notes, is designed to help field-workers, be they amateur or professional, to identify as many animals as possible in their samples. By so doing it is hoped that the more neglected animal groups, such as the tanaids, will receive the attention they truly deserve.

R. S. K. Barnes
Estuarine and Brackish-water
Sciences Association

Doris M. Kermack
The Linnean Society
of London

Introduction

The Tanaidacea is one of the orders of the crustacean class Malacostraca and, with six other orders, it is a member of the superorder Peracarida (McLaughlin, 1980). The approximately 800 living species occur almost exclusively in marine or brackish-water habitats, at a wide range of depths from the intertidal zone to the oceanic trenches (Wolff, 1956), and they have a worldwide distribution. In the North West Atlantic they may comprise up to 19% of the benthic macrofauna at 5000 m depth (Wolff, 1977). Tanaids are usually benthic and are tube-dwelling, burrowing or errant (free-living) in habit. Little is known of their ecology except in the cases of a few littoral species and those inhabiting some Scottish sea lochs (Holdich and Jones, 1983).

Until 1888 tanaids were included in other peracarid orders. Milne Edwards (1840) included them in the Isopoda but suggested that they constituted a group transitional to the Amphipoda. Dana (1852) placed them in a miscellaneous group of isopods which he called the Anisopoda, and Bate and Westwood (1868) included them with anthurid and gnathiid isopods in the 'Isopoda aberrantia'. Sars (1882) placed them in the Isopoda Chelifera and in 1888 Claus, retaining Dana's name, upgraded the Anisopoda to an independent order. Their separation as a distinct order is justified since only in this group of peracarids does the carapace cover, and invariably fuse with, the first two thoracic somites. Hansen (1895) suggested the name Tanaidacea which is now commonly used.

The systematics of the Tanaidacea is in a state of fluctuation, particularly at the family level. Identification is made difficult by the presence of sexual dimorphism, hermaphroditism, and the similarity of species within a genus (e.g. *Leptognathia*). Geographical variation in the shape of body segments and appendages can also occur. Useful general identification works are Sars (1896),* Norman and Scott (1906), Hansen (1913), Nierstrasz and Schuurmans Stekhoven (1930), and Holthuis (1956).

Until recently this order was divided into two suborders, the Monokonophora and the Dikonophora, on the basis of the form of the antennules, mandibles and genital cones (Lang, 1956). However, Sieg (1980a) has now included some fossil species in the classification and suggested that the division between the Monokonophora and Dikonophora is invalid. He

* Sars published the part of his *Crustacea of Norway* covering the Isopoda Chelifera in 1896; the Isopoda *sensu stricto*, however, were published in 1899 and sometimes the two parts are bound together and dated 1899 (Greve, personal communication).

has therefore divided the Tanaidacea into four new suborders, i.e. Anthracocaridomorpha (fossil), Apseudomorpha, Tanaidomorpha and Neotanaidomorpha.

The tanaids considered here are those which have been recorded from down to 200 m around the British Isles. Compared to the number of species recorded from the Northern Hemisphere, tanaids appear to be poorly represented in British waters with a total of only 27 species. However, this may well reflect a lack of collecting rather than a true lack of species. With more intensive surveys it seems likely than an increasing number of the species from neighbouring areas, such as Scandinavia, will also be found around the coasts of the British Isles. Tanaids are a moderately common component of the benthic fauna of the continental slope and regions such as the Rockall Trough, but these are outside the scope of the present *Synopsis*.

General structure

External anatomy

An apseudomorphan tanaid such as *Apseudes talpa* (Montagu) is typically dorsoventrally flattened with an anterior **cephalothorax** formed from the **cephalon** (head) and the first two **thoracomeres**. Dorsally and laterally the cephalothorax is covered by a **carapace** which may be produced anteriorly to form a **rostrum** (Fig. 1). Tanaidomorphan tanaids, e.g. *Leptognathia*, usually have a more cylindrical body (Fig. 22A). In all tanaids the lateral folds of the carapace form a **branchial chamber**. Behind the cephalothorax there are six free thoracomeres, or **pereonites**, forming the **pereon** (peraeon, pereion) (Fig. 1). Lateral extensions of these pereonites, the **epimera**, are often small and inconspicuous. The abdomen is composed of five somites, the **pleonites** – comprising the **pleon**, which may be fused in some genera – and a **pleotelson** formed from the sixth pleonite fused to the **telson** (Fig. 1). The dorsal and lateral surfaces of the integument sometimes bear grooves and setae but spines and tubercles are rare. The colour of tanaids is usually off-white or cream, sometimes with mottling due to **chromatophores**. Body length ranges from 1 mm to 25 mm, although most British specimens are between 1 mm and 6 mm.

The cephalothorax may bear compound **eyes** on small, usually immovable, anteriorly directed lobes (Fig. 1), although in some species, particularly those from the deep sea, eyes may be absent. The cephalothorax bears one pair of **antennules** (first antennae) and one pair of smaller **antennae** (second antennae) (Fig. 1). In the Apseudomorpha the antennule has two **flagella**, and the antenna one multi-articled flagellum and usually a characteristic scale-like, single-articled, **exopodite** (Fig. 2A, B). In the Tanaidomorpha and Neotanaidomorpha, however, the antennule and antenna each have only a single flagellum (Fig. 2C, D). The flagella of both antennules and antennae may bear setae and chemosensory **aesthetascs**, which are usually better developed in males.

Posterior to the insertion of these appendages, in mid-ventral line, lies the **epistome** (Fig. 1), which in some species is produced into a prominent anteriorly directed spine. The oral opening (mouth) is bordered anteriorly by the **labrum**, posteriorly by the bilobed **labium (paragnath)** and laterally by the **mandibles**. In the apseudomorphan family Apseudidae the lobes of the paragnath each bear an articulated projection anteriorly (Fig. 3H). The mandibles are asymmetrical (Fig. 3F, G). They bear an anteriorly directed **palp** and, on their inner surface, a distal **incisor process**, a **lacinia mobilis**

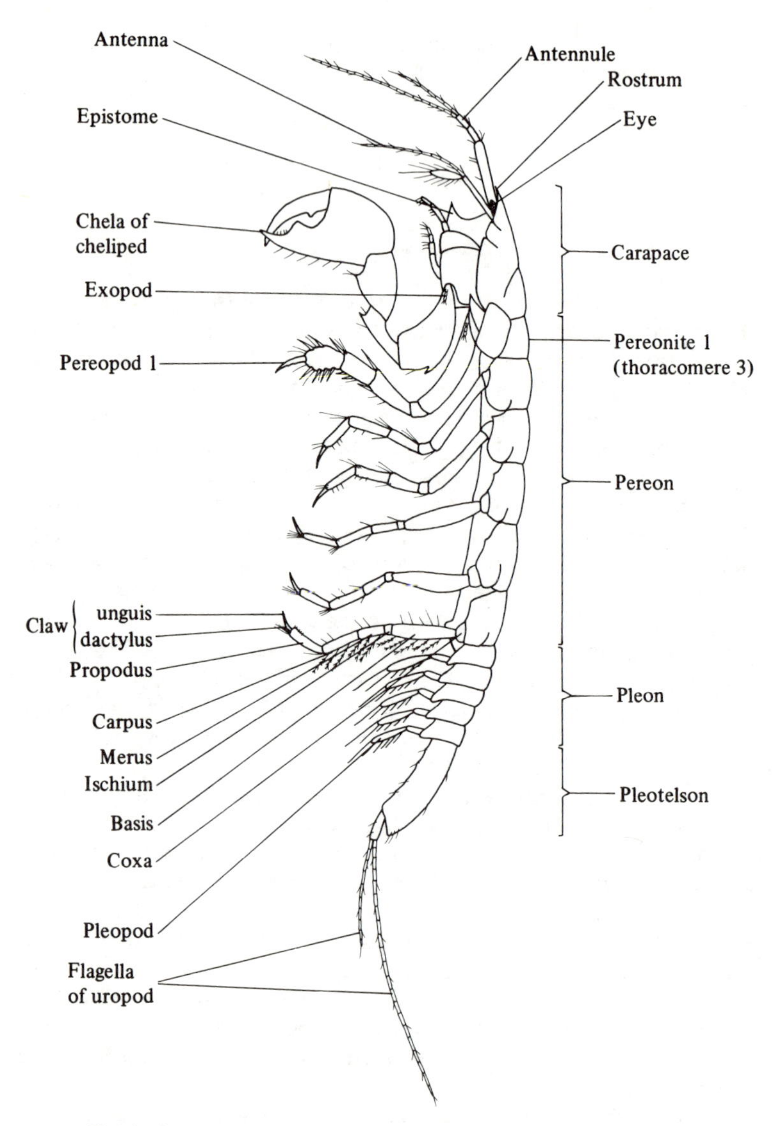

Fig. 1. Generalised apseudomorphan tanaid – based on *Apseudes*.

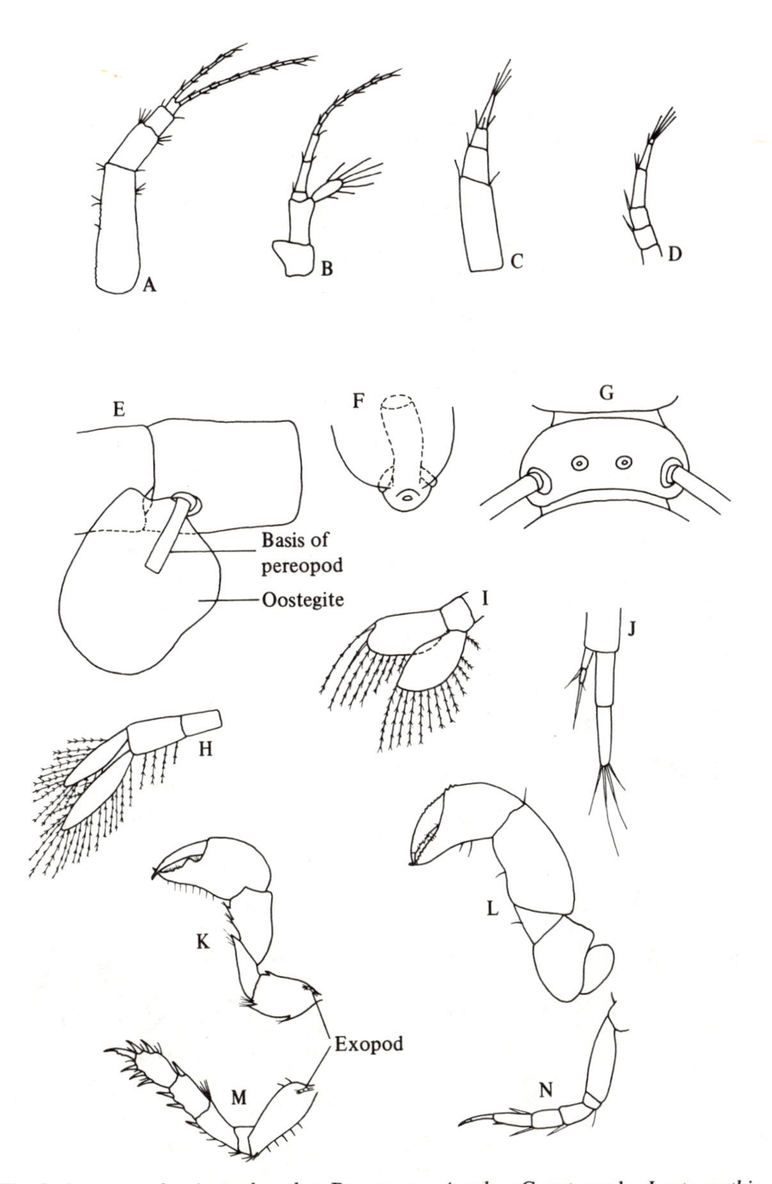

Fig. 2. A, antennule, *Apseudes talpa*; B, antenna, *A. talpa*; C, antennule, *Leptognathia brevimana*; D, antenna, *Leptognathia manca*; E, lateral view of tanaid showing position of an oostegite; F, female gonopore at base of oostegite and pereopod 4; G, genital cones on ventral surface of pereonite 6 of a male tanaidomorphan tanaid; H, pleopod, *Apseudes talpa*; I; pleopod, *Leptognathia gracilis*; J, uropod, *L. gracilis*; K, cheliped, *Apseudes talpa*; L. cheliped, *Leptognathia gracilis*; M, pereopod 1, *Apseudes talpa*; N, pereopod 1, *Leptognathia gracilis*.

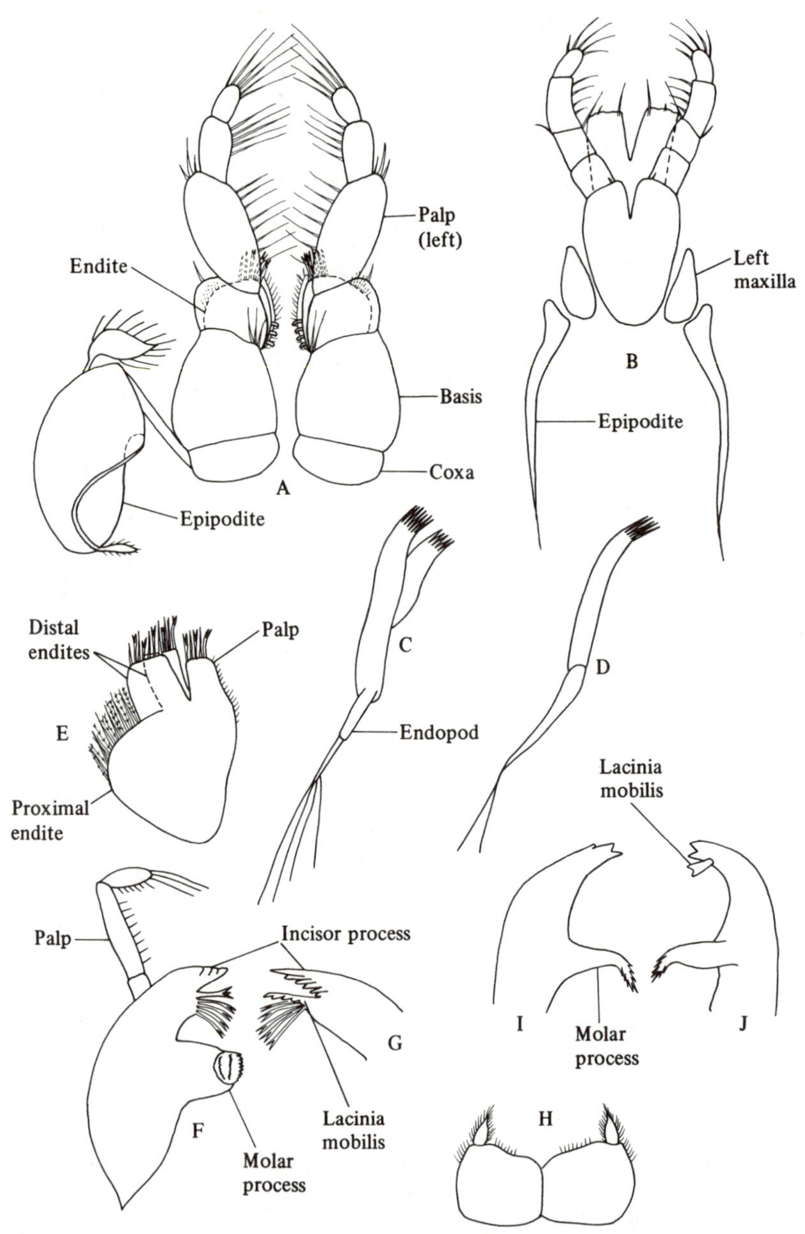

Fig. 3. A, maxillipeds, *Apseudes talpa*; B, maxillipeds and maxillae, *Leptognathia gracilis* female; C, maxillule, *Apseudes talpa*; D, maxillule, *Leptognathia gracilis*; E, maxilla, *Apseudes talpa*; F, right mandible, *A. talpa*; G, distal portion of left mandible, *A. talpa*; H, labium, *A. talpa*; I, left mandible *Leptognathia gracilis*; J, right mandible, *L. gracilis*.

which is often better developed on the left mandible, a row of forked setae, and a heavily thickened **molar process**. In other families these structures tend to be less well developed (Fig. 3I, J). In the Tanaidomorpha the epistome, labrum and mandibles have become shifted anteriorly, compared with the situation in the Apseudomorpha, to a position similar to that in the Isopoda and Amphipoda (Lauterbach, 1970). Behind the paragnath lie the **maxillules** (first maxillae), **maxillae** (second maxillae) and **maxillipeds**, the latter covering the other mouthparts (Fig. 3A, B). The maxillules have one or two (in Apseudomorpha) **endites** and an **endopod** (palp), bi-articled in apseudomorphans, which lies in the branchial chamber (Fig. 3C, D). The maxillae are structurally simple (Fig. 3E). In the apseudomorphans the **exite** is absent but two endites are present. The movable distal endite possesses two lobes and a palp – all well armed with stout setae. In *Apseudes* the proximal endite bears two rows of regularly arranged setae which are filtratory in function (Dennell, 1937). In suborders other than the Apseudomorpha the maxillae are usually vestigial. The maxillipeds (the first thoracic appendages) are usually well developed, frequently bearing a three- or four-articled palp and sometimes with a prominent **epipodite (epignath)** which extends into the branchial chamber as in the Cumacea (Fig. 3A, B). The left and right maxillipedal endites may be joined together by coupling hooks (Fig. 3A) and in Tanaidomorpha the two bases exhibit varying degrees of fusion (Fig. 3B). Males of some tanaidomorphan genera show a partial, sometimes even total, reduction of the mouthparts.

The second fused thoracomere bears the second pair of thoracic appendages, the **chelipeds**, and each of the six following free pereonites bears a pair of **pereopods** (walking legs) which are usually similar in form. A pereopod comprises seven articles except in members of the family Tanaidae where the ischium is absent. The proximal article, the **coxa**, is followed by the **basis** (usually the longest article), the **ischium, merus, carpus, propodus** and **dactylus** which terminates in an **unguis** (Fig. 1). The two latter articles may sometimes be fused to form a claw. The chelipeds are usually well developed and chelate in tanaids (Figs. 1, 2K, L), those of the male often being more robust. The propodus has a distal extension, the fixed finger, which with the movable dactylus, forms the chela. There has been some debate as to the naming of the pereopodal articles in tanaids, especially with regard to the cheliped which has only six articles. Sieg (1977) calls the articles of the cheliped: coxa, basis, merus, carpus, propodus and dactylus and these terms are used here. In some members of the Apseudidae the distal articles of the first pereopod are flattened to assist in digging or sometimes swimming (Figs. 1, 2M). These fossorial appendages are as long as the body in male *Sphyrapus*. Small exopods are usually present on the bases of the chelipeds and first pereopods in the Apseudomorpha (Figs. 1, 2K, M). In *Apseudes* these **exopods** assist in producing a water current through the branchial chamber. In one family, the Kalliapseudidae, they are also present on the last three pairs of pereopods during early post-marsupial stages. There are indications that the ancestors of the Tanaidacea had exopods on all pereopods thus suggesting links with the Mysidacea and Cumacea.

As with other peracarid crustaceans, ovigerous female tanaids possess a **marsupium** (brood pouch), formed from a number of **oostegites**, within which the eggs undergo their development. The lamellate oostegites arise from the coxae on the inner side of certain pereopods (Fig. 2E, F). Initially they appear as small structures which reach their full size after the maturation moult, and overlap or fuse in the midline. However, in the Neotanaidae and Tanaidae oostegites are fully formed after the maturation moult but remain enclosed in spherical cuticular sheaths which open after the moult to form the marsupium (Gardiner, 1975a). Some tanaids are able to open and close the oostegites in order to ventilate the brood. Four pairs of oostegites occur in most tanaids, although in some families, e.g. Tanaidae, there is only a single pair arising from the fourth pereopods. This pair may form either a single marsupium or two separate **brood sacs** (Lang, 1960). In certain tanaids there may also be a pair of small oostegites arising from the chelipeds, but they do not form part of the marsupium (Gardiner, 1973).

The female **gonopores** are situated on the coxae of the fourth pereopods and are difficult to distinguish (Fig. 2F). The male gonopores occur on the **sternum** of pereonite 6. In the Apseudomorpha they are usually situated on a single **genital cone** and in most Tanaidomorpha on a pair of cones (Fig. 2G). However, some confusion may arise when trying to sex specimens on the basis of male gonopores as individuals with oostegites may sometimes also possess genital cones.

Each pleonite usually bears a pair of natatory **pleopods** (Fig. 1), consisting of a coxa, a basis, and two flattened branches – endopodite and exopodite – bearing setae (Fig. 2H, I). In tanaids the pleopods do not appear to have a respiratory function and are absent in adult females of some species. When present they may be used in swimming, and in helping to produce a ventilatory current through the tube of tube-dwelling species. Terminally the pleotelson bears a pair of uni- or biramous **uropods** (Fig. 2J). In some genera, e.g. *Apseudes*, they are very long and multi-articled (Fig. 1).

In the foregoing account the terminology of Sieg (1977), rather than that of Wolff (1956) or Gardiner (1975a), is used for subdividing the tanaid body. It should be noted that the numbering of the pereonites in this and other *Synopses* in the series concerning peracarid crustaceans, i.e. isopods (Naylor, 1972) and cumaceans (Jones, 1976), differs. All three orders have eight thoracomeres. In the Isopoda the first is fused to the head and bears a pair of maxillipeds. The remaining seven free thoracomeres are termed the pereon and bear pereopods 1-7. In the Cumacea the first three thoracomeres are fused to the head and each bears a pair of maxillipeds. The remaining five free thoracomeres form the pereon and bear pereopods 1–5. In the Tanaidacea the first two thoracomeres are fused to the head and bear the maxillipeds and chelipeds respectively. The pereon is formed from the remaining six free thoracomeres and bears pereopods 1–6. Previous authors have taken the tanaid chelipeds as the first pereopods and consequently the first free pereonite, according to their terminology, is known as pereonite 2. McLaughlin (1980), however, has listed the free pereonites as 3–8 and yet numbered their corresponding appendages as pereopods 2–7 in order to try and standardise terminology within these groups of crustaceans.

Internal anatomy

The main organ in the body cavity of a tanaid is the **alimentary tract**. This is composed of an ectodermal **foregut**, an **hepatopancreas** and a long ectodermal **hindgut** which opens on the ventral side of the pleotelson between a pair of anal flaps. The foregut lies in the carapace region and is divided into a short muscular **oesophagus** and a 'stomach' region; the latter is subdivided into a trituratory **cardiac** (chewing) portion and a filtratory **pyloric** region. The 'stomach' of *Apseudes*, at least, is more complex than in genera such as *Heterotanais* and *Tanaissus* (Siewing, 1954). The hepatopancreas arises at the junction of the pyloric foregut and the hindgut, and is composed of two or three pairs of **caeca** which lie alongside the hindgut in the pereonal cavity.

The nervous system is usually well developed and consists of a **brain** connected by **circum-oesophageal connectives** to a **nerve cord** which lies beneath the alimentary tract. In *Apseudes* the ganglia are widely separated but in *Heterotanais* and *Tanaissus* those in the carapace region and pleon lie closer together than those in the pereon (Siewing, 1954).

The circulatory system consists of a **heart**, anterior, lateral, and posterior arteries, and blood sinuses. The tubular heart lies dorsal to the alimentary tract and in *Apseudes* and *Heterotanais*, at least, it extends the length of the pereon, although in *Tanaissus* it is much shorter. The heart lies in a **pericardial sinus** with a muscular **pericardial membrane**. At least two pairs of **ostia** are present. Blood from the appendages flows back to the pericardial sinus in membrane-lined channels. Anteriorly the artery extending into the carapace region widens to form an accessory heart (**cor frontale**) which supplies the brain, antennae and labrum with blood. The number of lateral arteries varies between taxa, there being one pair in most species but two pairs in *Heterotanais* and three pairs in *Apseudes*. The blood is largely oxygenated in an extensive capillary network associated with the inner walls of the carapace.

Excretion is by means of a pair of glands which open at the base of the maxillae. Rudimentary **antennal glands** have also been found in some species.

In sexually mature tanaids the reproductive organs fill much of the pereonal cavity not occupied by the alimentary tract. In females the **ovaries** are long and tubular and open through an **oviduct** on either side of the fourth pereonal sternite. The **testes** of male tanaids are much shorter, their **vasa deferentia** either fuse to form a single **seminal vesicle**, or remain separate to form a pair of seminal vesicles. In either case the openings of the vasa deferentia are paired and are situated on a single median **genital cone** or a pair of cones, on the sixth pereonal sternite. **Hermaphroditic** species of tanaid possess both ovaries and testes simultaneously.

Tegumental glands occur in the pereonal cavity and in tube-dwelling species such as *Tanais* and *Heterotanais* they secrete a silk-like substance. This is used in the construction of the animal's tube, and passes from the glands along long ducts which open on the dactyli of certain pereopods.

Biology

Life history

Marsupial development

As in other peracarid crustaceans, tanaids undergo their early development within the confines of a marsupium, on the ventral surface of an ovigerous female, and are not usually liberated until they have gained most of their appendages. However, in *Tanais dulongii* (Audouin) embryos have occasionally been found developing separately from the parent but still within the confines of the female's tube. This also appears to occur in *Typhlotanais aequiremis* (Lilljeborg).

Scholl (1963) has described the embryology of *Heterotanais oerstedi* (Krøyer). The fertilised egg develops epimorphically into an embryo which shows marked dorsal curvature, i.e. the dorsal tip of the pleotelson touches the dorsal part of the carapace (Fig. 4A). This gradually straightens out and develops into the first **manca stage** (Manca I) (Fig. 4B), and then into a second, more mobile manca stage (Manca II) (Fig. 4C). This stage has partially formed sixth pereopods and eventually leaves the marsupium (Buckle Ramirez, 1965). There is some controversy as to whether there are two or three manca stages during tanaid development (Messing, 1981). *Apseudes talpa*, from British coasts, appears to follow a similar developmental sequence to *H. oerstedi.* Marsupial development in *H. oerstedi* apparently proceeds without moult, i.e. according to the descriptions given (Scholl, 1963; Buckle Ramirez, 1965) there is no shedding of various membranes to release characteristic stages as occurs in isopods (Holdich, 1968).

Post-marsupial development and reproduction

The most thorough study of post-marsupial development is that made by Buckle Ramirez (1965) on the complex life history of the tube-dwelling species, *Heterotanais oerstedi* (Fig. 4D). As previously mentioned Manca II leaves the marsupium. This is followed by a juvenile stage (**Neutrum** I of some workers) which has fully formed appendages, and at the next moult these juveniles develop either into preparatory males or preparatory females. The latter may be gonochoristic or sequentially hermaphroditic. Crustaceans are generally not hermaphroditic, and if hermaphroditism occurs in malacostracans it is usually of the protandrous variety (Charniaux-Cotton, 1960). In tanaids, however, many different patterns are found. In some taxa simultaneous, protandrous and protogynous hermaphrodites have been recorded. Lang (1953) in a study of *Apseudes hermaphroditicus* Lang (this has now

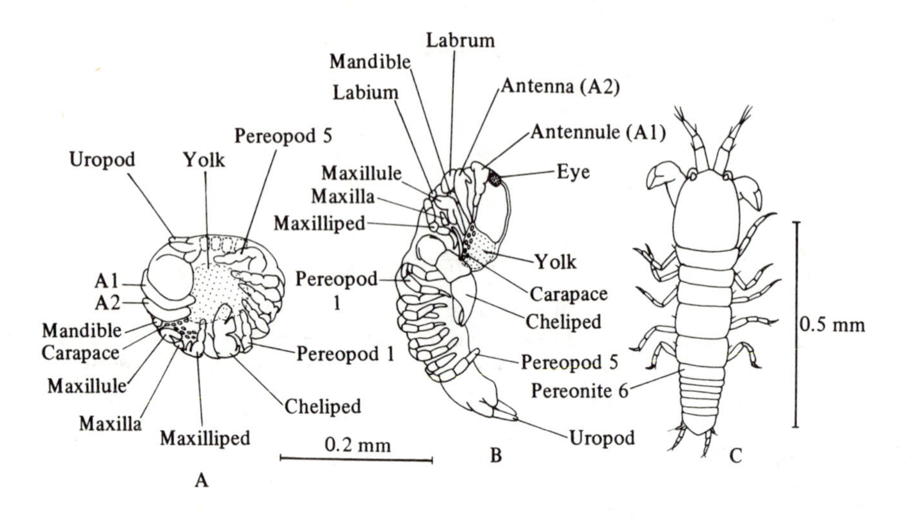

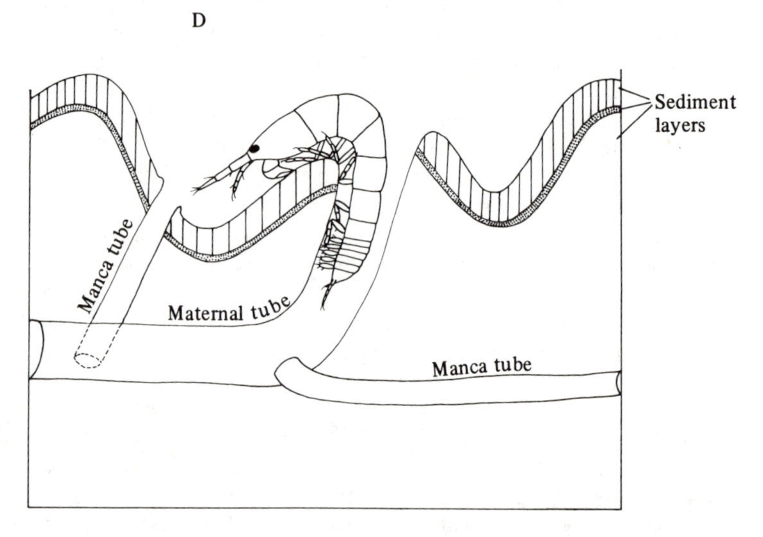

Fig. 4. A, Embryo of *Heterotanais oerstedi* (after Scholl, 1963); B, Manca I of *H. oerstedi* (after Scholl, 1963); C, Manca II of *H. oerstedi* (after Buckle Ramirez, 1965); D, *H. oerstedi* female emerging from a tube constructed in a thin layer of sediment. Manca tubes are shown arising from the maternal tube.

been synonymised with *Apseudes spectabilis* Studer) recorded specimens which possessed both mature ova and a male genital cone. He found that in those specimens with mature eggs in the oviducts the male gonads were filled with sperm, and concluded that self-fertilisation was possible. There are a number of recorded cases of tanaids with characteristically male chelipeds, but with small oostegites or eggs in a marsupium, and ovaries and testes both present.

H. oerstedi has been termed a potential hermaphrodite by Buckle Ramirez (1965) because, in the laboratory at least, environmental conditions are important in determining the sex of an individual. Early stages raised in isolation result in either gonochoristic or protogynous females. In the presence of females, however, juveniles may develop into primary copulatory males after passing through the preparatory stage. It has been suggested that the males produce a pheromone which prevents juveniles and protogynous females from developing into males. In addition to primary copulatory males, secondary copulatory males can arise from three sources, i.e. from copulatory females which have not been fertilised and which have resorbed their eggs (Type A); from females which have had at least one brood (Type B); and from females which have moulted to an intermediate female stage after releasing a brood (Type C). The existence of two forms of male in this species has led to some taxonomic confusion since primary and secondary males differ in size and cheliped shape (Fig. 13B, C). In the past it was believed that the primary copulatory male constituted a different species, *H. gurneyi* Norman. The male stage is terminal in the life cycle of *H. oerstedi* but after release of her brood a female moults to either an intermediate female with small oostegites or a secondary copulatory male (Type B). The intermediate female may either moult to a secondary female and produce a second brood, or moult to a secondary copulatory male (Type C). The stages mentioned above are summarised in the diagram below.

Prior to mating in *H. oerstedi* a copulatory male enters the tube of a copulatory female and there is a long period of courtship. Male and female come to lie with the ventral sides adjacent and sperm is deposited in the marsupium as the female opens her oostegites. After closing the marsupium the female lays up to sixteen eggs. She then drives the male from the tube, closes up the tube ends, and incubates her brood. After leaving the marsupium, the mancas cut their way through the maternal tube and build side branches (Fig. 4D). Marchand (1977) has also examined the life cycle of *H. oerstedi* and has given additional information on the population dynamics of this species.

Sieg (1978a) has suggested that development is similar in all tanaids, generally involving three manca stages, a juvenile stage, preparatory males and females, and primary copulatory males and females. However, as mentioned previously, some workers believe that there are only two manca stages (Messing, 1981). In the normal type of development, copulatory females stay as females and may enter a second copulatory stage via an

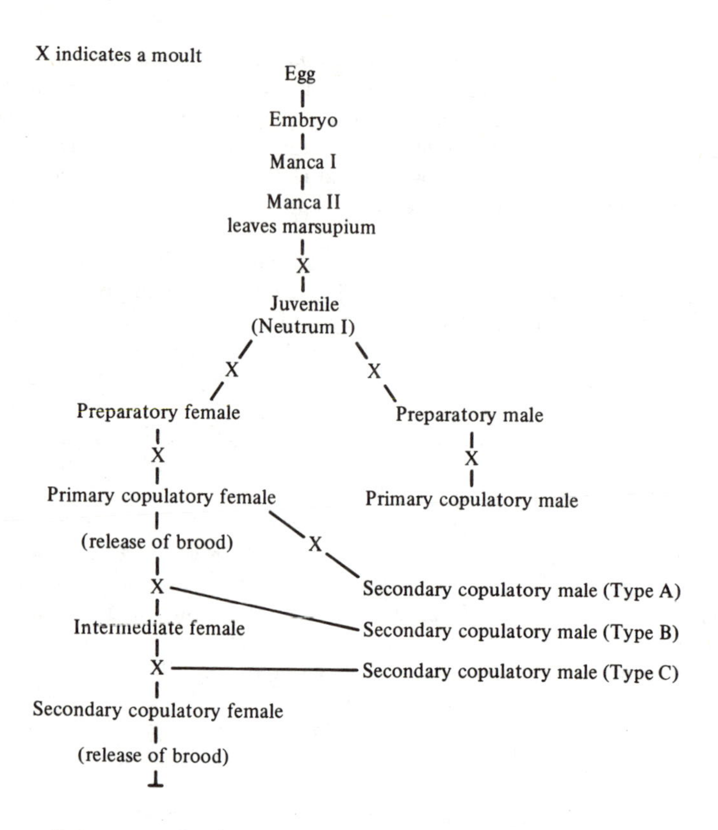

intermediate stage. In the potential hermaphrodite type of development, both primary and secondary copulatory females may change into secondary copulatory males.

The reproductive cycles of those shallow-water tanaids that have been studied exhibit a seasonal pattern of breeding which varies with latitude. In an *Apseudes latreillii* (Milne Edwards) population in south-west France, Salvat (1967) found breeding females from March to December, with juveniles being especially abundant in November and December. In contrast, in a population of *H. oerstedi* in north Germany, where winter temperatures are much lower, Buckle Ramirez (1965) found that the young were produced in early July and that mancas produced later died with the onset of winter.

Sexual dimorphism appears to be the rule amongst the Tanaidacea. The males of some species have more highly developed pleopods and greater development of sensory setae on the antennules and antennae than the females. Males of some genera, e.g. *Heterotanais* and *Leptochelia* also possess much larger eyes than females. The reason for these differences is not known, but they may be concerned with mate location. Buckle Ramirez (1965) reports that *H. oerstedi* males use their chelae to open the tubes of the female prior to mating, but that the act of copulation can take place successfully when males have their chelipeds amputated. The chelipeds of the male are larger than those of the female – a common feature in a number of other genera.

Feeding

Most tanaids are raptorial feeders consuming detritus and its associated micro-organisms. Large food masses are seized by the chelipeds and maxillipeds and are abraded by the stout setal armature of the maxillules, maxillae and maxillipeds. Medium sized food particles are also retained on the setules of the maxillipedal setae. These particles are transferred towards the mouth by the setae on the proximal endite of the maxilla and are then removed by the maxillules and mandibles. The mouthparts of an apseudomorphan tanaid are shown *in situ* in Figs. 5A and 6B and those of a tanaidomorphan in Fig. 5B.

In the case of the tube-dwelling species it is possible that diatoms, nematodes and detritus within the tube of the female may provide a source of nourishment for the mancas (Marchand, 1977).

A common feature of male tanaidomorphan and neotanaidomorphan tanaids is a reduction of the mouthparts, and in the Neotanaidae the anal opening is also permanently closed (Gardiner, 1975a). These morphological features may be correlated with the observation that the gut of copulatory males in some species is completely empty. Buckle Ramirez (1965) states that male *H. oerstedi* do not feed after the maturation moult.

In some members of the Apseudidae raptorial feeding is supplemented by filter feeding and, accordingly, they possess better developed branchial chambers, maxillae, maxillipedal epipodites, and exopods on the chelipeds and first pair of pereopods. Dennell (1937) has described in detail the complex filter feeding mechanism employed by *Apseudes talpa* to obtain a small fraction of its food. Particles are trapped by maxillary filter setae, comprising part of a filter chamber formed by the maxillae and maxillipeds (Figs. 5A, 6B). The suction which causes water to enter the filter chamber is initiated by rhythmic movements of the maxillipeds and by the respiratory current (see p. 16). The filtrate of small particles is removed from the maxillary setae by dorsal setae on the maxillipedal endite and passed towards the mouth.

Respiration

In tanaids the main respiratory surface is the integument of the carapace wall facing the branchial chamber. In *Tanais dulongii* at least, the maxillipedal epipodites also have a respiratory function (Lauterbach, 1970, as *T. cavolinii*). A comparison of two British littoral tanaids, *Apseudes talpa* and *Tanais dulongii*, shows that their branchial chambers differ structurally and in the direction of current flow. In *T. dulongii* (Fig. 5B) the large base of the cheliped and the folded edge of the carapace form the posterior wall of the chamber. Anteriorly the chamber is closed by the fusion of the carapace with the lateral wall of the head. The maxillipedal epipodites activate the respiratory currents and cause water to pass in an antero-lateral to postero-dorsal direction through the chamber (Fig. 5B). Food particles are not filtered from the water during its passage through the chamber. In contrast, the branchial chambers of *A. talpa* (Figs. 5A, 6B) are closed posteriorly by the swollen bases of the chelipeds, laterally by the maxillules and maxillae, and anteriorly by the swollen mandibular bases. The epipodite of the maxilliped is hemispherical, with a concavity fitting against the swollen base of the cheliped. Upward movement of the epipodites causes water to flow into the branchial chamber at the ventro-posterior margins of the carapace (the inhalent opening) and also through a passage between the bases of the maxillules and maxillae that leads firstly into the filter chamber. Downward movement of the epipodite causes water to exit near the base of the maxilliped, all other exits being closed at this time (Fig. 5A). Small particles of food are filtered by the maxillary setae at the entrance to the filter chamber (see p. 15). The endopods of the maxillule act as cleaning devices for the respiratory surfaces and the exopods on the bases of the chelipeds and first pereopods aid in the movement of water towards the inhalent opening. Unlike *T. dulongii* the respiratory current of *A. talpa* is from posterior to anterior (Fig. 5A).

Gamble (1970) has compared the anaerobic survival of *Parasinelobus chevreuxi* (as *Tanais chevreuxi*) with that of two *Corophium* species (Amphipoda). The tanaid has a fairly high resistance to anaerobic conditions which probably reflects its semi-sessile crevice-dwelling mode of life. Unlike *Corophium*, *P. chevreuxi* tends not to move away from its habitat when low oxygen concentrations occur.

Ecology

Very little is known about the ecology of tanaids, especially those living offshore. Most species are benthic, although *Kalliapseudes* spp. have been found in the plankton. Some tanaids, e.g. *Tanais grimaldii* Dollfus, can swim very fast for short periods by beating their pleopods.

Although tanaids are mainly a marine group some have penetrated freshwater habitats. The best known example is that of the euryhaline species *Sinelobus stanfordi* Richardson (formerly *Tanais stanfordi*). This can live in waters of salinity ranging from 0–52‰. On the Galapagos Islands it lives in both hypersaline and freshwater lakes where it constructs tubes on soft and hard substrata, on algae, rushes and mangroves, and sponges (Gardiner, 1975b). *S. stanfordi* also tolerates a wide range of temperature, inhabiting tropical and temperate regions along the coasts of the Atlantic, Pacific and

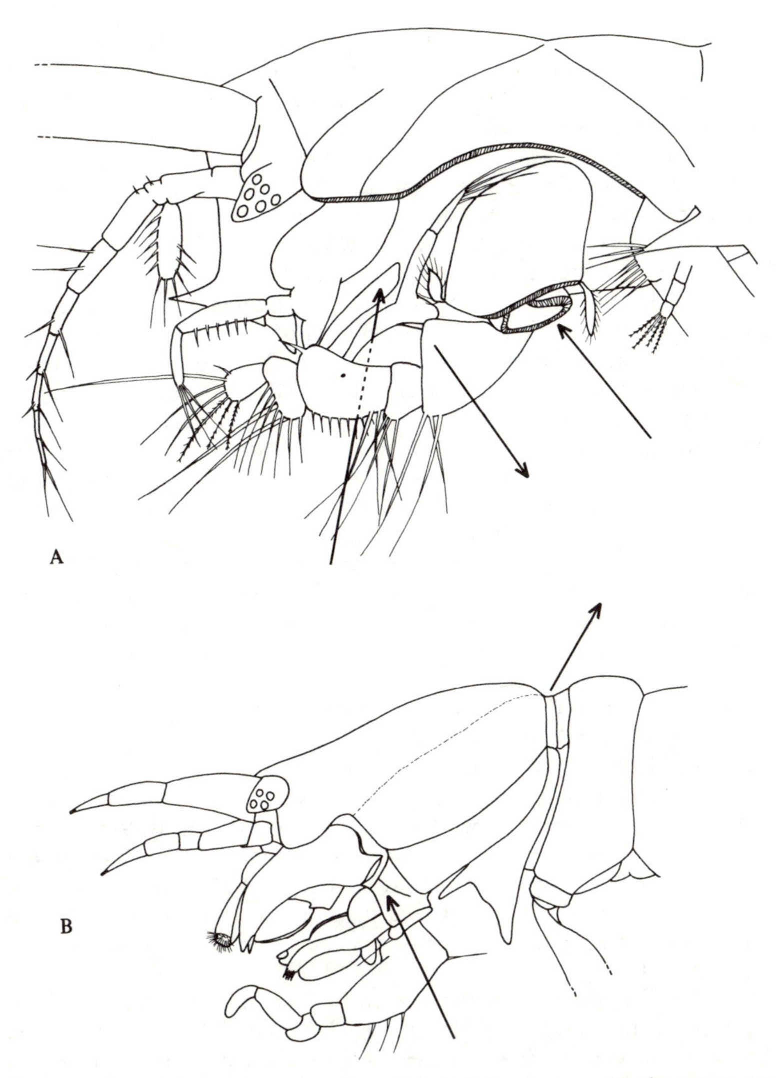

Fig. 5. A, lateral view of *Apseudes talpa* with left cheliped removed and carapace cut away to show mouthparts (after Dennell, 1937); B, lateral view of *Tanais dulongii* with left cheliped removed (after Lauterbach, 1970). Arrows indicate inhalent and exhalent currents.

Indian Oceans in lagoons, lakes, estuaries and rivers. *Heterotanais oerstedi* is the only British species which has penetrated freshwater.

Two American tanaids having unusual habits are *Hexapleomera robustus* (Moore) and *Pagurapseudes* spp. The former inhabits minute tubes in crevices between the scales of a turtle's carapace, the pleon of the latter is adapted to fit into small gastropod mollusc shells in a similar manner to the abdomen of hermit crabs (Messing, 1979).

Many tanaids construct tubes using a sticky 'cement' to which particles of sand and detritus adhere. This cement is produced from tegumental glands in the pereonal cavity and passes through tubes to emerge from the tips of the dactyli of the anterior pereopods. The number of pereopods involved in this process depends on the species. *Heterotanais oerstedi* uses the anterior three pairs of pereopods to spin its tube and *Tanais dulongii* uses at least the first pair of pereopods. Females of the latter species will construct a new tube from sand grains and detritus in under six hours if they are removed from their original tubes. Normally the tubes of tanaids are open at both ends and the pleopods are used to create a through current. Ovigerous females may, however, seal off the ends of their tubes during the incubatory period. When feeding, tube-dwelling tanaids may emerge completely or only partially from their tubes (Fig. 4D). In *T. dulongii* the tubes can reach 10 cm in length and often form a complex containing many individuals within a crevice (Holdich and Jones, 1983). When disturbed the tanaids retreat rapidly down their respective tubes.

Little is known about the predators of tanaids. It seems possible that fish are amongst the main predators. Most species of bottom feeding fish from the Rockall Trough have been found to contain tanaids in their stomach contents (J. Mauchline, personal communication).

Some tanaid genera are found at a wide variety of depths. Wolff (1956) states that in *Apseudes* some twenty species live at depths from 1400–6800 m, seventeen species live in shallower water, twelve in the littoral zone, and four in shallow water of low salinity. On the other hand species of the genus *Neotanais* usually live in deep water on the surface of oceanic oozes (Gardiner, 1975a).

Most of the offshore species recorded from around the British Isles have been found in association with sand, mud or gravel. Some of the deeper species also occur at depths below 200 m. On rocky shores a number of species occur in muddy crevices, amongst barnacles, in coralline rock pools and kelp holdfasts. On muddy shores tanaids are sometimes found in waterlogged wood and amongst *Enteromorpha*. Few species have been recorded from intertidal sand although *Tanaissus lilljeborgi* Stebbing is occasionally recorded as being common in this habitat. The ecology and distribution of British tanaids has been summarised by Holdich and Jones (1983).

Many tanaid species, notably the fossorial apseudomorphans such as *Apseudes talpa* and *A. latreillii*, are able to bury themselves quickly in sediment using their antennules and antennae to loosen the sediment and the first pereopods to excavate actively. Although *A. talpa* is usually recorded littorally in the British Isles, Salvat (1967) has found that its density is sometimes eight times higher at 3–9 cm depth compared with that at 0–3 cm depth in the sublittoral sands of south-west France.

Collection, preservation and examination

Tanaids are difficult to see in the field and it is often better to collect a portion of a likely habitat and sort it in the laboratory. Both sorting and examination may be facilitated by staining the sample in Rose Bengal.

Sublittoral species associated with particulate substrata are best collected with a grab, anchor dredge or epibenthic sled. The tanaids may be separated from the sediment by elutriation using a large stream of water running at low velocity (Gage, 1972). Water running off the surface should be passed through at least a 420 μm sieve to retain the smaller tanaids. In addition, many of the tanaids may float to the surface, as they are displaced by the water jet, and can be removed before entering the sieve. Another method is to use a diver-operated suction sampler with a fine mesh bag. Many of the 'fine sieves' used by ecologists have a 1 mm mesh and consequently small crustaceans, such as tanaids, are lost from the sample.

On rocky shores tanaids such as *Apseudes talpa* can be found between mean tide level (MTL) and low water spring level (LWS) in deep muddy crevices. Here it is best to use a hammer and chisel to prise open the crevices. *Tanais dulongii* is often found in slate crevices and associated with rock pools containing coralline algae and *Cladophora*. As this species inhabits tubes it is often difficult to determine whether the tubes are of annelid or tanaid origin in the field.

To extract tanaids from algae and crevice debris, samples should be placed in a shallow dish with water containing a few drops of 20% Biofix (Gerrard) or 10% formalin. This irritates the fauna often causing the cryptic species to emerge. As with the elutriation method small specimens may get caught in the surface film and are easily visible as light is reflected off their shiny cuticles.

Tanaids may be killed in 10% neutral formalin or Biofix and then transferred to 70% alcohol for storage.

A dissecting microscope with a high-power objective is needed to examine most tanaids. Small specimens and appendages may be more closely examined after staining in CMC-9A (Masters Chemical Co., USA) which not only stains the specimen but clears it and can be used as a mountant for permanent microscope slides. In order to prevent specimens becoming dehydrated under the microscope they can be transferred from alcohol into neat glycerol.

Classification

(extant suborders and families only)

Order Tanaidacea Hansen, 1895

Suborder Apseudomorpha Sieg, 1980
- Family Apseudellidae Gutu, 1972
- Family Apseudidae Leach, 1814
- Family Cirratodactylidae Gardiner, 1973
- Family Gigantapseudidae Kudinova-Pasternak, 1978
- Family Kalliapseudidae Lang, 1956
- Family Leviapseudidae Sieg, 1980
- Family Metapseudidae Lang, 1970
- Family Pagurapseudidae Lang, 1970
- Family Tanapseudidae Bacescu, 1978

Suborder Neotanaidomorpha Sieg, 1980
- Family Neotanaidae Lang, 1956

Suborder Tanaidomorpha Sieg, 1980
- Family Tanaidae Dana, 1849
- Family Paratanaidae Lang, 1949
- Family Leptognathiidae Sieg, 1976
- Family Pseudotanaidae Sieg, 1976
- Family Agathotanaidae Lang, 1971
- Family Nototanaidae Sieg, 1976
- Family Anarthruridae Lang, 1971

Systematic part

In the classification given above, which is based on Sieg (1980a), superfamilies have been omitted. In the Apseudomorpha all the extant species are included in the superfamily Apseudoida Leach, 1814, and this is synonymous with Lang's Monokonophora (see Introduction). Lang's Dikonophora is synonymous with the suborders Tanaidomorpha and Neotanaidomorpha.

Apseudomorphan families are poorly represented in British waters with only the Apseudidae having been recorded. However, all seven of the families in the Tanaidomorpha are present. Twenty-seven species of tanaid are now known from down to depths of 200 m around the British Isles (in addition a new species of Nototanaidae is currently being described by the authors). Prior to the present study Norman's (1899) list of British tanaids was the most comprehensive, although he also included species outside our depth range.

Few keys to the Tanaidacea have been published although Sieg and Winn (1978) have recently produced one to many of the families listed above. Sieg (1980b) has extensively revised the family Tanaidae, and has changed the generic or specific names of some well known tanaids, e.g. *Tanais cavolinii, Tanais chevreuxi* and *Tanais stanfordi.*

The main distinguishing features of the Apseudomorpha are as follows: antennule with two flagella; antenna usually with scale-like exopod; mandible with palp; cheliped and pereopod 1 often with small exopods; four or five pairs of oostegites; single genital cone in male. By contrast the Tanaidomorpha have: antennule with one flagellum; antenna without exopod; mandible without palp; cheliped and pereopod 1 without exopods; one to five pairs of oostegites; usually two genital cones in males. The apseudomorphan genus *Sphyrapus* is unusual in that it does not have a scale-like exopod on the antenna and the inner flagellum on the antennule consists of only one article. However, it is unlikely to be confused with any tanaidomorphan species.

Key to British families

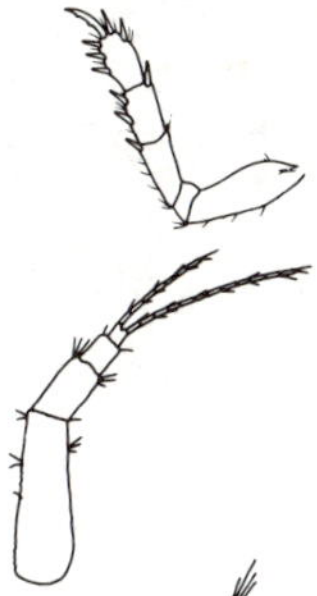

1. Pereopod 1 powerfully built, fossorial, propodus flattened. Antennules biflagellate. Male with single genital cone on pereonite 6.......... .... Family Apseudidae (p. 27)

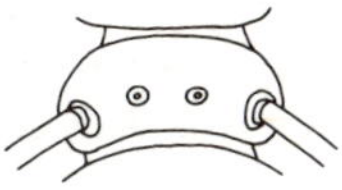

   Pereopod 1 similar to pereopods 2–6. Antennules uniflagellate. Male with two genital cones on pereonite 6 .... **2**

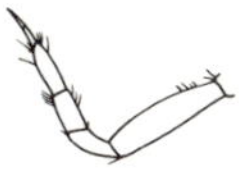

2. Pereopods lacking ischium; three pairs of pleopods present; uropods uniramous .... Family Tanaidae (p. 38)

   Pereopods with ischium; when present, five pairs of pleopods; uropods usually biramous... **3**

3. Chelipeds without coxa; uropods strongly degenerate; pleopods absent in female .... Family Agathotanaidae (p. 78)

   Chelipeds with coxa; uropods not degenerate **4**

4. Coxa of cheliped articulating with mediolateral margin of basis; six pleonites **5**

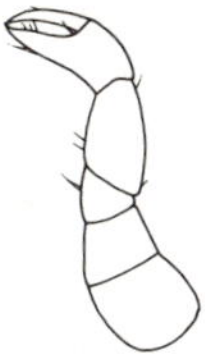

   Coxa of cheliped large, articulating with proximal margin of basis; one or six pleonites .... Family Anarthruridae (p. 88)

5. Maxilliped bases (or vestiges in male) not fused medially; endopodite of pleopod with distinct seta proximally on inner margin .... Family Paratanaidae (p. 44)

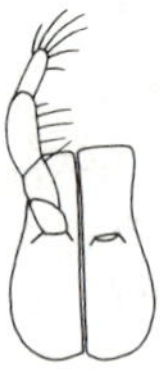

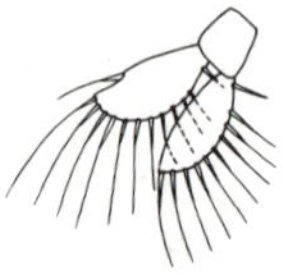

Maxilliped bases fused medially to a varying degree; endopodite of pleopod not as above... **6**

6. Bases of maxillipeds totally fused; endites also completely fused or almost so........ .... **7**

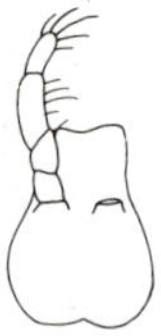

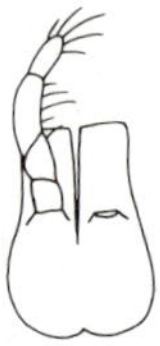

Bases of maxillipeds not completely fused..... ....Family Leptognathiidae (p. 50)

7. Antennule three-articled in female, seven-articled in male; when present marsupium formed from one pair of oostegites; body short and broad .... Family Pseudotanaidae (p. 80)

Antennule four-articled in female, six-articled in male; when present marsupium formed from four pairs of oostegites; body elongated .... Family Nototanaidae (*Tanaissus*) (p. 86)

The British species of tanaids

SUBORDER APSEUDOMORPHA
FAMILY APSEUDIDAE
GENUS APSEUDES Leach, 1814
A. talpa (Montagu, 1808)
A. latreillii (Milne Edwards, 1828)
A. spinosus (M. Sars, 1858)
A. grossimanus Norman, 1886
GENUS SPHYRAPUS Norman and Stebbing, 1886
S. malleolus Norman and Stebbing, 1886
SUBORDER TANAIDOMORPHA
FAMILY TANAIDAE
GENUS TANAIS Latreille, 1831
T. dulongii (Audouin, 1826)
GENUS PARASINELOBUS Sieg, 1980
P. chevreuxi (Dollfus, 1898)
FAMILY PARATANAIDAE
GENUS HETEROTANAIS Sars, 1882
H. oerstedi (Krøyer, 1842)
GENUS LEPTOCHELIA Dana, 1849
L. savignyi (Krøyer, 1842)
FAMILY LEPTOGNATHIIDAE
GENUS PSEUDOPARATANAIS Lang, 1973
P. batei (Sars, 1882)
GENUS TYPHLOTANAIS Sars, 1882
T. tenuicornis Sars, 1882
T. brevicornis (Lilljeborg, 1864)
T. microcheles Sars, 1882
T. aequiremis (Lilljeborg, 1864)
T. pulcher Hansen, 1913
GENUS TANAOPSIS Sars, 1896
T. graciloides (Lilljeborg, 1864)
GENUS LEPTOGNATHIA Sars, 1882
L. brevimana (Lilljeborg, 1864)
L. breviremis (Lilljeborg, 1864)
L. gracilis (Krøyer, 1842)
L. manca Sars, 1882
L. paramanca Lang, 1958
L. filiformis (Lilljeborg, 1864)
?*L. rigida* (Bate and Westwood, 1868)
FAMILY AGATHOTANAIDAE
GENUS AGATHOTANAIS Hansen, 1913
A. ingolfi Hansen, 1913

FAMILY PSEUDOTANAIDAE

GENUS PSEUDOTANAIS Sars, 1882

P. forcipatus (Lilljeborg, 1864)

P. jonesi Sieg, 1977

FAMILY NOTOTANAIDAE

GENUS TANAISSUS Norman and Scott, 1906

T. lilljeborgi Stebbing, 1891

FAMILY ANARTHRURIDAE

GENUS ANARTHRURA Sars, 1882

A. simplex Sars, 1882

Description of families, genera and species

Family APSEUDIDAE Leach, 1814

Body dorsoventrally flattened and tapering posteriorly. Carapace well developed with prominent rostrum, ocular lobes distinct, eyes may be well developed or absent. Pereonite 1 closely applied to the posterior edge of the carapace. Remaining pereonites and all pleonites sharply demarcated. Antennules biflagellate; antennae uniflagellate, sometimes with a scale-like exopodite. Mouthparts always well formed. Mandibles with a three-articled palp. Chelipeds often sexually dimorphic. Pereopod 1 with distal articles flattened and propodus with many spines, the whole limb being powerfully built and adapted for digging. Cheliped and pereopod 1 usually with a small exopod. Pleopods well developed. Uropods biflagellate and with very long endopod. Although this family contains many genera only *Apseudes* and *Sphyrapus* have been recorded from British waters.

Key to British genera and species of *Apseudidae*

1. 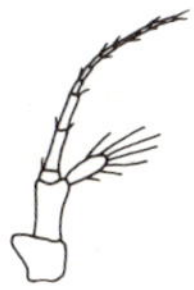 Antenna with setose scale arising from second peduncular article .... *Apseudes* **2**

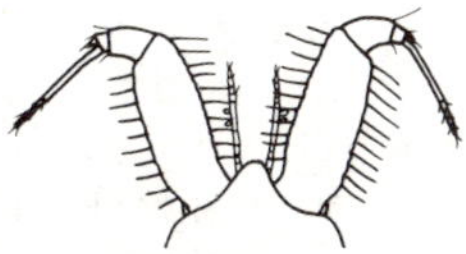

Antenna without scale ....*Sphyrapus* (*Sphyrapus malleolus*) (p. 36)

2. Rostrum parallel sided with rounded apex and spiniform tip (Fig. 7A), epistomal spine absent .... *A. latreillii* (p. 30)

Rostrum not as above, epistomal spine present **3**

3. Rostrum broadly triangular with minutely serrated margins, ventrally keeled, depressed at apex (Fig. 6A) .... *A. talpa* (p. 28)

Rostrum long, acute apical projection **4**

4. Rostrum deflexed with broad base and upwardly turned lateral lobes (Fig. 8A) .... *A spinosus* (p. 32)

Rostrum tridentate, central spine longer than others and very acute (Fig. 9A) .... *A. grossimanus* (p. 34)

Genus *APSEUDES* Leach, 1814

Cephalothorax flattened and carapace distinctly sculptured with prominent rostrum. Ocular lobes generally well developed. Antennule similar in males and females with two unequal, multi-articled flagella. Antenna with one multi-articled flagellum and a single-articled scale. Epistomal spine present or absent. Coxa of pereopod 1 produced as an anteriorly directed process. Pereonites with or without medio-ventral spines. Pleotelson lacking acute, dorsally deflected process.

Apseudes talpa (Montagu, 1808)

(Figs. 2A, B, H, K, M; 3A, C, E–H; 5A; 6)

Cancer (Gammarus) talpa Montagu, 1808
Eupheus talpa Desmarest, 1825
Apseudes talpa: Leach, 1814; Bate and Westwood, 1868
Apseudes hibernicus Walker, 1897
non Apseudes talpa: Lilljeborg, 1864; Sars, 1886; Nierstrasz and Schuurmans Stekhoven, 1930

Rostrum short, triangular, with serrated margins and deflected apex; base showing varying degree of indentation (Fig. 6A). Eyes present. Epistome and sternite of thoracomere 2 each with a medial spine (Fig. 6B). Antero-lateral margins of pereonites with short, rounded processes. Pleonal epimera prominent, apically truncate and usually highly setose (Fig 6A). Pleotelson same length as pleon. Chelipeds not markedly sexually dimorphic but those of male slightly more robust. In adults there is a prominent tooth on the inner margin of the chelipedal propodus (Fig. 6A). *A. talpa* is usually of a porcelain white colour although this is often partially masked by dirt in the grooves on the carapace and attached to the setae. Body length 5–8 mm (excluding uropods).

A. talpa appears to be capable of hermaphroditism as specimens from the British Isles include adults with partially or fully formed oostegites but with a male genital cone, and individuals with male chelae and oostegites.

A. talpa is mainly found on the mid to lower shores associated with mud or sand under stones and in crevices, and from *Laminaria* holdfasts and coralline algae. It also occurs in the sublittoral zone. Recorded from the south and west coasts of Britain, the Channel Islands, the Isles of Scilly, the Isle of Man and the east and west coasts of Ireland. There are also records from Dutch, French and north African coasts. Sars (1886) figured an '*A. talpa*' from the Mediterranean which possessed prominent ventral pereonal spines. These are never seen in British *A. talpa* although a small tubercle may be present on some pereonites. Norman and Stebbing (1886) have suggested

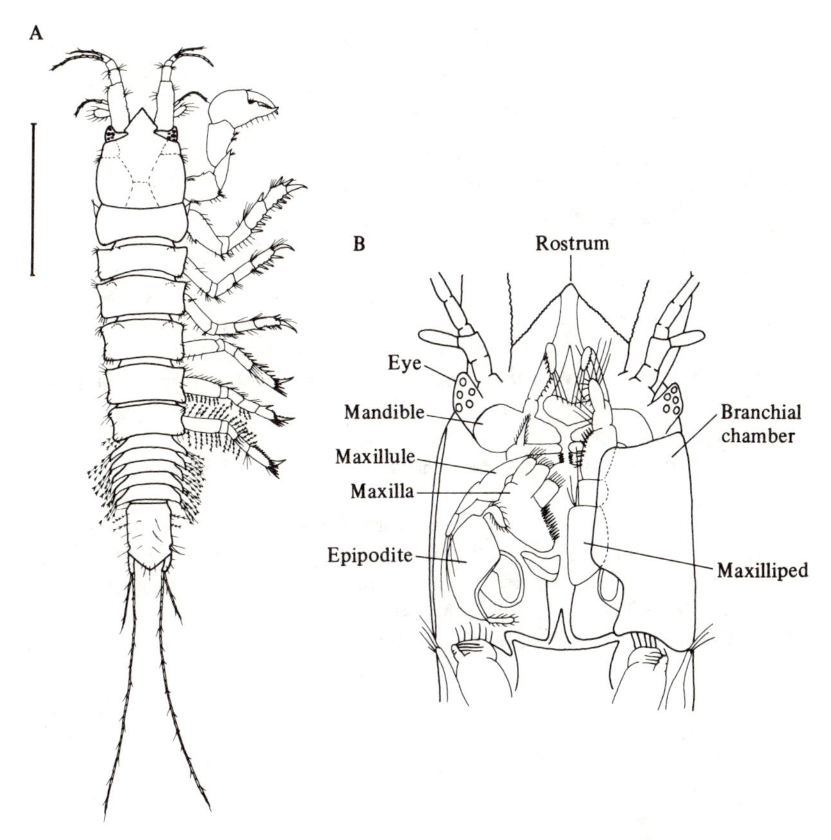

Fig. 6. *Apseudes talpa.* A, dorsal view (scale bar 2 mm); B, ventral view with right-hand portion of carapace removed to show mouthparts (after Dennell, 1937).

that Sars' specimens were in fact *A. spinosus. A. hibernicus* Walker from Ireland was based on tanaids lacking ventral pereonal spines. Walker (1897) had presumably compared his specimens with those of Sars (1886) and found them to differ in this character – they were in fact *A. talpa* (Montagu).

Apseudes latreillii (Milne Edwards, 1828)

(Fig. 7)

Rhoea Latreillii Milne Edwards, 1828
Apseudes Latreillii: Bate and Westwood, 1868
Apseudes Latreillii: Sars, 1886
non Apseudes Latreillii: Claus, 1884; 1888.

Rostrum distinct, short but acutely spiniform and deflexed at apex, sides parallel (Fig. 7A). Eyes present. Epistomal spine absent. Medio-ventral pleonal spines present. Antero-lateral margins of pereonites not produced. Pleonal epimera posteriorly deflected at apices, each apex with a spine (Fig. 7A). Pleotelson shorter than pleon. The chelipeds are markedly sexually dimorphic (Fig. 7B, C). In the male they are robust with a prominent tooth on the inner margins of the propodus and dactylus. In the female the chela is more slender and without teeth. Body length 4–7 mm (excluding uropods).

A. latreillii is found on the mid to lower shore under stones, in rock crevices, and in muddy gravel. It also occurs in kelp holdfasts, coralline algae and *Zostera* roots. Recorded from southern and north-eastern coasts of England, the Channel Islands, the Isles of Scilly, and eastern Scotland. Records also exist for the French and Italian coasts.

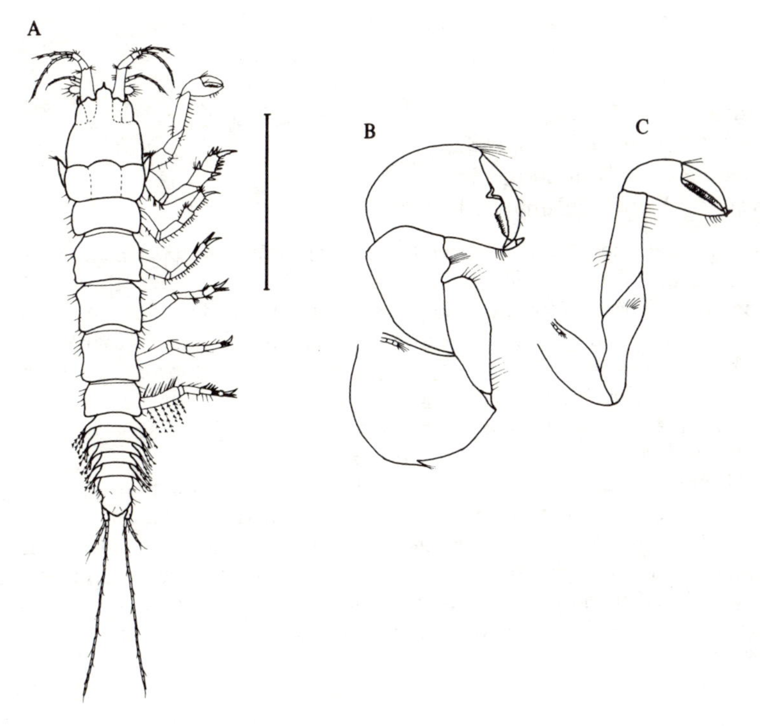

Fig. 7. *Apseudes latreillii.* A, female, dorsal view (scale bar 2 mm); B, cheliped, male; C, cheliped, female.

Apseudes spinosus (M. Sars, 1858)

(Fig. 8)

Rhoea spinosa M. Sars, 1858
Apseudes spinosus: Sars, 1882; 1886
Apseudes talpa: Lilljeborg, 1864
Apseudes Koehleri Bonnier, 1896

Rostrum well developed, deflexed, sides bulbous and upturned, apex produced into a prominent projection (Fig. 8A). Eyes absent. Epistomal spine present. Medio-ventral spine on each pereonite and pleonite. Antero-lateral margins of pereonites 2–6 each produced into a prominent triangular, pointed process (Fig. 8A). Pleonal epimera well developed. Pleotelson longer than pleon. Chelipeds of male and female similar, those of male being more robust, both with tooth on inner margin of propodus (Fig. 8B, C). Body length 8–12 mm (excluding uropods).

This species has been recorded from the northern North Sea, from the Fladen Station (58°20′N, 00°30′E) at 140 m (McIntyre, 1961), and further north at 110 m (58°40′N, 1°15′E) and 131 m (59°17′N, 00°00′E). It occurs commonly in Norwegian fjords (54–270 m), and has also been recorded off Iceland, Sweden and Denmark, south-south-west Ireland and in the Bay of Biscay. It is found in diverse bottom conditions at depths from 18 m to 1300 m.

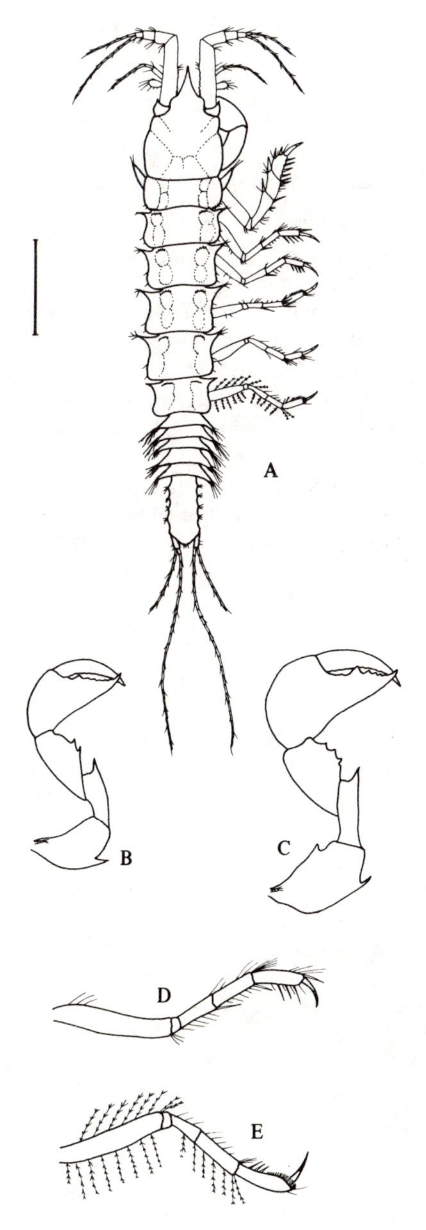

Fig. 8. *Apseudes spinosus*. A, dorsal view (scale bar 2 mm); B, cheliped, female; C, cheliped, male; D, pereopod 3; E, pereopod 6.

Apseudes grossimanus Norman*

(Fig. 9)

Apseudes grossimanus Norman, 1870 (*nomen nudum*)
Apseudes grossimanus Norman, 1881 (*nomen nudum*)
Apseudes grossimanus Norman in Norman and Stebbing, 1886

Rostrum markedly tridentate with central spine nearly as long as basal antennal peduncle article (Fig 9A). Prominent, acute processes on antero-lateral margins of carapace and pereonites 2–6 (Fig. 9A). Medio-ventral spine present on all pereonites and pleonites. Pleonal epimera pointed, pleotelson narrow and longer than pleon. Antennules, antennae and pereopods long and slender (Fig. 9A). Maxillipeds with strong spine on inner surface of the basis (Fig. 9D). Chelipeds markedly sexually dimorphic (Fig. 9B, C), that of male being more robust and with a strong tooth on the inner margin of the propodus. Body length up to 17 mm (excluding uropods).

A. grossimanus is normally a deep-water species but it has been recorded once off south-west Ireland at a depth of 162 m. In deeper water it has been found in the Bay of Biscay, off Portugal and the northern and south-western coasts of Africa.

* This species was not described until 1886, although it was mentioned by Norman (1870, 1881). Norman and Stebbing (1886) attribute it to Norman.

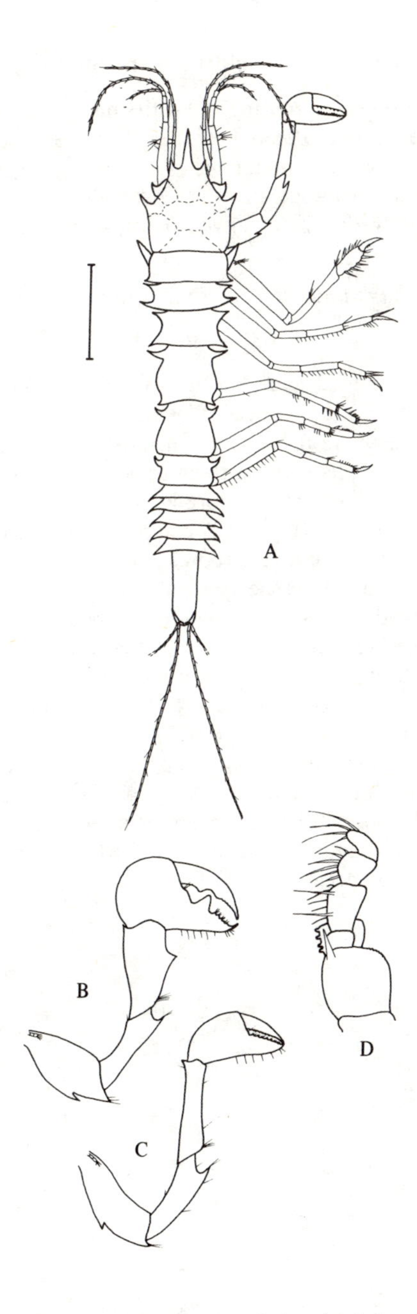

Fig. 9. *Apseudes grossimanus*. A, female, dorsal view (scale bar 2 mm); B, cheliped male; C, cheliped, female; D, maxilliped.

Genus *SPHYRAPUS* Norman and Stebbing, 1886

Carapace without distinct sculpturing and with simple rounded rostrum. Eyes absent, ocular lobes minute. Pereonite 1 very closely applied to the posterior margin of the carapace. Antennules with inner flagellum sometimes rudimentary. Antenna without scale-like exopodite. Chelipeds with large chela. Pereopod 1 very long, especially in males, with distal articles flattened.

Sphyrapus malleolus Norman and Stebbing, 1886

(Fig. 10)

Rostrum short with rounded apex (Fig. 10A, C). Pereonites irregular in width and length, without lateral processes. Pleonites without prominent epimera except on pleonite 2 where they are characteristically wing-like (Fig. 10A). Pleotelson shorter than pleon and with an upturned spiniform apical process (Fig. 10B). Antennules sexually dimorphic with first peduncular article of male being short and stout (Fig. 10C) compared with that of female (Fig. 10A). Inner flagellum of antennule represented by only one article. Antennae with two pear-shaped vesicles on outer margin of the fourth article. Chelipeds sexually dimorphic but well developed in both sexes. Chela malleolate in male (Fig. 10D) and ovate in female (Fig. 10A). Pereopod 1 much longer than other pereopods; pereopod 3 shorter than any except the last; pereopods 4–6 with propodus and dactylus surrounded by serrate spines. Body length ≃5 mm (excluding uropods).

Although *S. malleolus* is normally a deep-water species there is one record from south of Rockall (Norman and Stebbing, 1886) which brings it within the scope of this Synopsis. It has also been recorded from deeper water south of Rockall, from the Bay of Biscay, and off Portugal and Greenland, along with other *Sphyrapus* species.

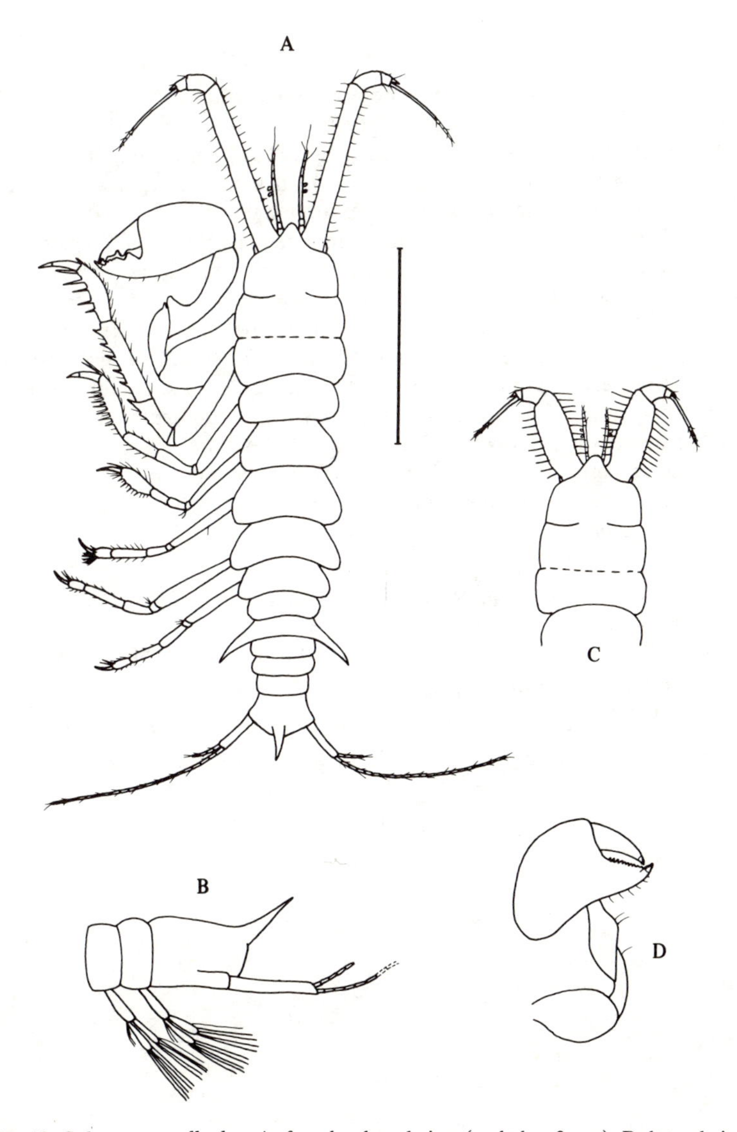

Fig. 10. *Sphyrapus malleolus*. A, female, dorsal view (scale bar 2 mm); B, lateral view of posterior end of female; C, anterior end of male; D, cheliped, male.

Family TANAIDAE Dana, 1849

Cephalothorax usually truncated anteriorly. Ocular lobes distinct, eyes well developed. Pleon of three to five pleonites. Antennule with three-articled peduncle which may bear a one- to two-articled flagellum. Antenna six–eight articled. Maxillipeds not fused and with small but distinct coxae. Labium with an inner and outer lobe, the latter sometimes bearing small lobes or spines at the antero-lateral corners. Pereopods lacking ischium, pairs 4–6 with a well developed claw (fused dactylus and unguis). Three pairs of pleopods present. Uropods uniramous with a basis and a two- to many-articled endopodite. The female possesses only one pair of oostegites which arise from the coxae of pereopods 4. These form one or two sac-like protrusions. Sexual dimorphism is slight and is represented by differences in body proportions and in the form of the cheliped, antennule and antenna. The mouthparts are not reduced in the male. There are only two species of Tanaidae represented in British waters, *Tanais dulongii* and *Parasinelobus chevreuxi*, these may be distinguished from each other by the form of the uropods, antennae and labium (Figs, 11, 12).

Key to British species of Tanaidae

1. Uropods three-articled (Fig. 11D). Outer lobe of labium with antero-lateral projections (Fig. 11G) .... *Tanais dulongii* (p. 39)

Uropods four-articled (Fig. 12E). Outer lobe of labium without antero-lateral projections (Fig. 12D) .... *Parasinelobus chevreuxi* (p. 42)

Genus *TANAIS* Latreille, 1831

Tanais Audouin and Milne Edwards, 1829 (*nomen nudum*)
Tanais Milne Edwards, 1830 (*nomen nudum*)

Pleon of four pleonites, the first two of which bear a strong, semicircular row of erect setae dorsally. Antennule four-articled; antennae seven-articled (Fig. 11F), distal two articles small. Outer lobe of labium with antero-lateral projections (Fig. 11G).

Tanais dulongii (Audouin, 1826)

(Fig. 5B; 11)

Gammarus Dulongii Audouin, 1826
Tanais Dulongii: Milne Edwards, 1838, 1840
Tanais Cavolinii Milne Edwards, 1840
Tanais tomentosus Krøyer, 1842
Tanais vittatus Lilljeborg, 1864
Tanais hirticaudatus Norman, 1899
Tanais cavolinii: Lauterbach, 1970; *auct.*
Tanais dulongii: Sieg, 1980b

Carapace often with distinctive grey mottling. Pereonites with rounded margins, pereonite 1 much shorter in length than others (Fig. 11A). Some variation has been found in the relative proportions of the body in both sexes of this species. When mature, however, males tend to be more robust than females. Lacinia mobilis of left mandible well developed (Fig. 11C), that of right mandible reduced (Fig. 11B). Chelipeds variable in form – in some specimens a tooth-like process is seen on the inner margin of both the propodus and dactylus, in others one or both of these processes may be absent. Cheliped often rather stronger in male. Pereopods 4–6 similar; dactylus and unguis fused to form a claw which bears a row of comb-like setae and also small setose tufts on its proximal margin (Fig. 11E). Uropods with three-articled endopodite (Fig. 11D). Body length 3–5 mm.

Until recently *T. dulongii* was believed to be distinct from *T. cavolinii* Milne Edwards. However, Sieg (1980b) suggested that in the original figure of *Gammarus dulongii* the two rows of erect setae on the dorsal surface of the pleon had been accidentally omitted. This view is supported by the works of several other authors, e.g. Lucas (1849), whose *T. dulongii* is figured with these rows of setae. Hence *T. cavolinii*, which differed from *T. dulongii* only by its possession of the rows of erect setae on the pleon, is now considered a synonym of the latter species. Sieg (1980b) lists a very extensive synonomy.

T. dulongii (recorded as *T. cavolinii*) is a common species in the littoral and shallow sublittoral of the British Isles, where it inhabits self-constructed tubes of sand grains, mud and detritus in shallow pools, amongst clusters of barnacles and in rock crevices. It is often found in association with filamentous and coralline algae, and may also occur in burrows in wood formed by other invertebrates. There are records for *T. dulongii* (as *T. cavolinii*) from the south, west and north-east coasts of England; the north, east and west coasts of Scotland; the south-east and west coasts of Ireland; the Shetland Islands and Jersey. On a world-wide basis it has been found between northern Norway and the Mediterranean, from the east coasts of North America and South America, and from south-west Australia.

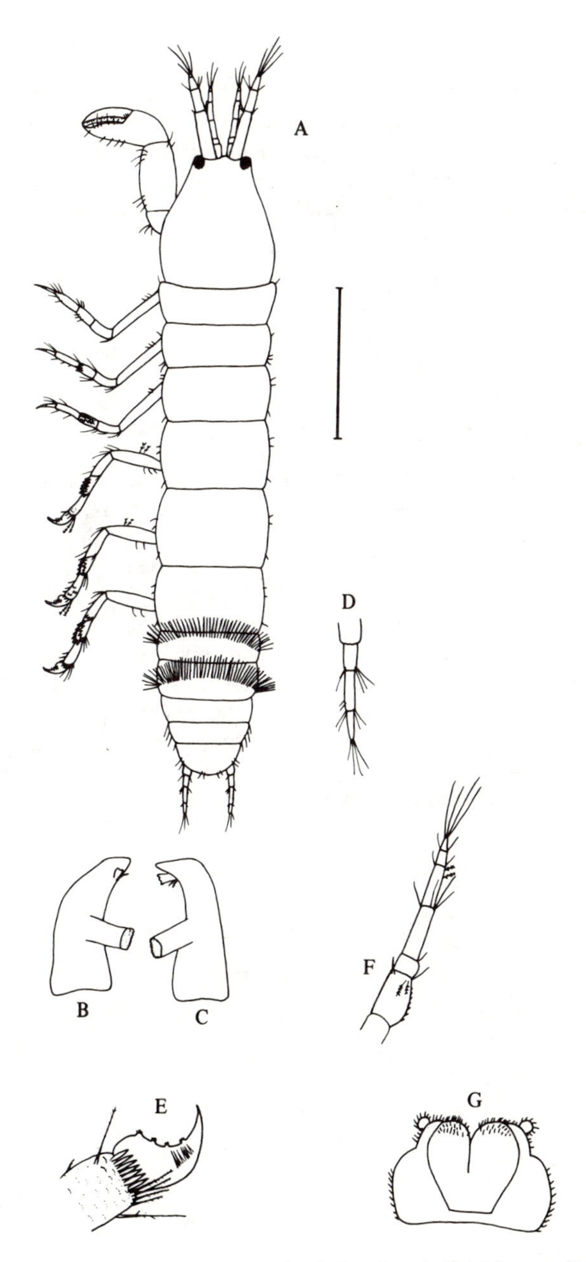

Fig. 11. *Tanais dulongii.* A, dorsal view (scale bar 1 mm); B, right mandible; C, left mandible; D, uropod; E, distal article of pereopod 6; F, antenna; G, labium.

Genus *PARASINELOBUS* Sieg, 1980

Pleon of four pleonites the first two of which bear a strong, semicircular row of erect setae dorsally. Antennule four-articled, last article bearing many setae and a few aesthetascs (Fig. 12A). Antenna six-articled, articles 2 and 4 with a semicircle of setae distally (Fig. 12B). Outer lobes of labium lacking antero-lateral projections (Fig. 12D).

Parasinelobus chevreuxi (Dollfus, 1898)

(Fig. 12)

Tanais chevreuxi Dollfus, 1898
Tanais chevreuxi: Gamble, 1970; *auct.*
Parasinelobus chevreuxi: Sieg, 1980b

Pereonites 1–3 of similar size. Lacinia mobilis of left mandible strong, of right mandible reduced. Cheliped often stronger in male than in female, and with dactylus rather more curved. Claws of pereopods 4–6 lacking marginal setose tufts seen in *T. dulongii*. Uropods four-articled (Fig. 12E). Body length 4–6 mm.

P. chevreuxi is usually found in the outer to middle region of intertidal crevices, where it inhabits self-constructed tubes of coarse sediment. The distribution of this species is rather restricted compared with that of *T. dulongii*. It has been recorded (as *Tanais chevreuxi*) from west Ireland, south-west and north Wales, and south Devon, where it is common at several localities (Marine Biological Association, 1957). Records also exist for the Atlantic coasts of France, Spain and Morocco.

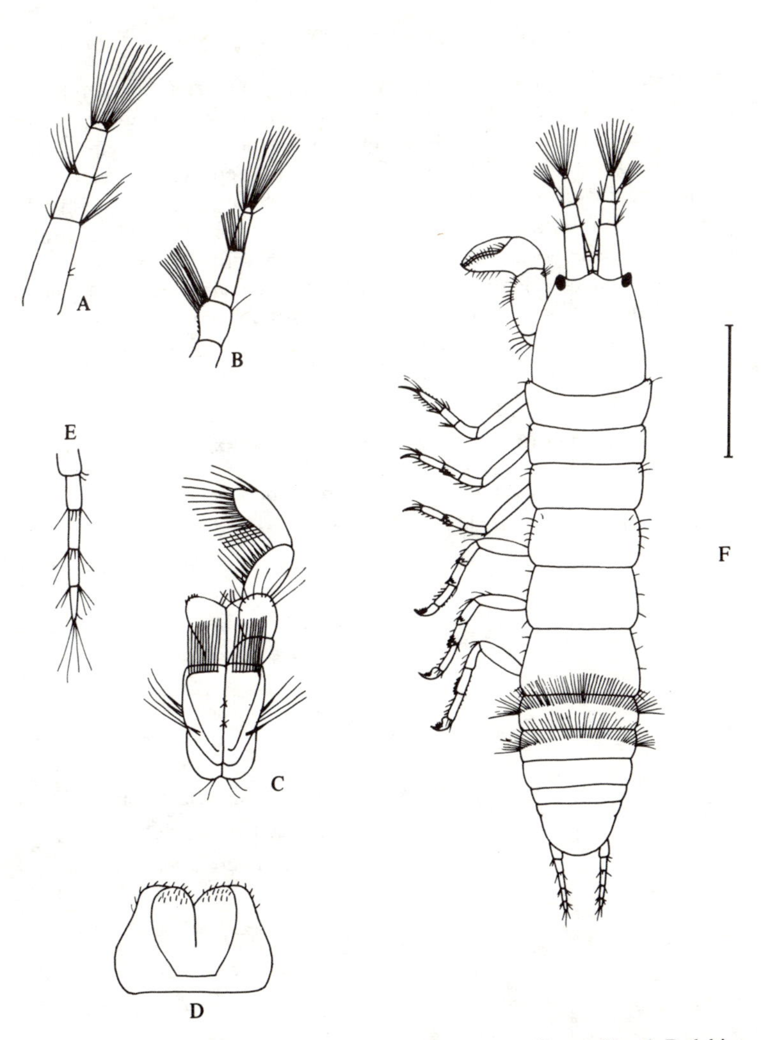

Fig. 12. *Parasinelobus chevreuxi.* A, antennule; B, antenna; C, maxilliped; D, labium; E, uropod. F, dorsal view (scale bar 1 mm).

Family PARATANAIDAE Lang, 1949

British species of this family are strongly sexually dimorphic. Antennule of female three- or four-articled, that of male with two- or three-articled peduncle and a multi-articled flagellum. Antenna six- or seven-articled. Maxillipeds not fused medially. Mouthparts of male entirely reduced except for slight remains of maxilliped. Palp of maxillule in female single-articled with two terminal setae. Coxa of cheliped situated behind proximal projection of basis. Pereopods with ischium, pairs 4–6 with dactylus and unguis fused to form a claw. Pleon with five pleonites each with a pair of biramous pleopods. Endopodite of pleopods with a distinctly separate seta proximally on inner margin. Uropods biramous. Marsupium formed from four pairs of oostegites.

Key to British species of Paratanaidae

Antenna with strong ventral spine on second article. Exopodite of uropod single-articled .... *Leptochelia savignyi* (p. 48)

Antenna without ventral spine on second article. Exopodite of uropod two-articled .... *Heterotanais oerstedi* (p. 46)

Genus *HETEROTANAIS* Sars, 1882

Antennule of female three-articled and without flagellum, that of male with three-articled peduncle and two-articled flagellum. Antenna with six articles. Mandible with cylindrical molar process and serrated chewing surfaces. Endite of maxillule bearing nine long spines, maxillipedal palp three-articled in male, four-articled in female. Chelipeds of female normal, those of male very large with a caudal extension at the end of the carpus and with a downwardly deflected propodus (Fig. 13B, C). Bases of pereopods 4–6 much thicker than those of 1–3. Exopodite of uropods two-articled (articulation sometimes indistinct), endopodite four-articled.

Heterotanais oerstedi (Krøyer, 1842)

(Figs. 4; 13)

Tanais Ørstedi Krøyer, 1842
Heterotanais Ørstedi: Sars, 1882
Heterotanais oerstedi: Nierstrasz and Schuurmans Stekhoven, 1930
Heterotanais gurneyi Norman, 1906
Heterotanais gurneyi: Hammond, 1974

Markedly sexually dimorphic. Eyes prominent in both sexes, especially in the male. Cephalothorax of male exceptionally elongated in form, that of female normal in appearance (Fig. 13A, D). Antenna with strong dorsal spine at distal corner of second and third articles (Fig. 13E). Diagnosis otherwise as for genus. Body length ≃2 mm.

This tube-dwelling species is unusual amongst British tanaids in that it exhibits tolerance to a wide range of salinites from freshwater to estuarine and fully marine conditions, preferring sheltered places with silt-covered vegetation and a depth not exceeding approximately 10 m. *H. oerstedi* builds a tube using the chelipeds and anterior three pairs of pereopods. The tubes may be stuck to plant matcrial, solid surfaces, or in mud (Fig. 4D). This species has a complex life history details of which have been given previously (pp. 11–15). It may appear abundantly in some localities for one year and virtually die out the next. The presence of more than one type of male has given rise to some taxonomic confusion (p. 13), with differences between the males being most obvious in the form of the chelipeds (Fig. 13B, C).

H. oerstedi has been recorded from marine, estuarine and fluvial locations on the south and east coasts of England, and from south Wales. Worldwide it has been found on the coasts of Norway, Sweden, Denmark, northern France and in the Kiel Canal, Germany.

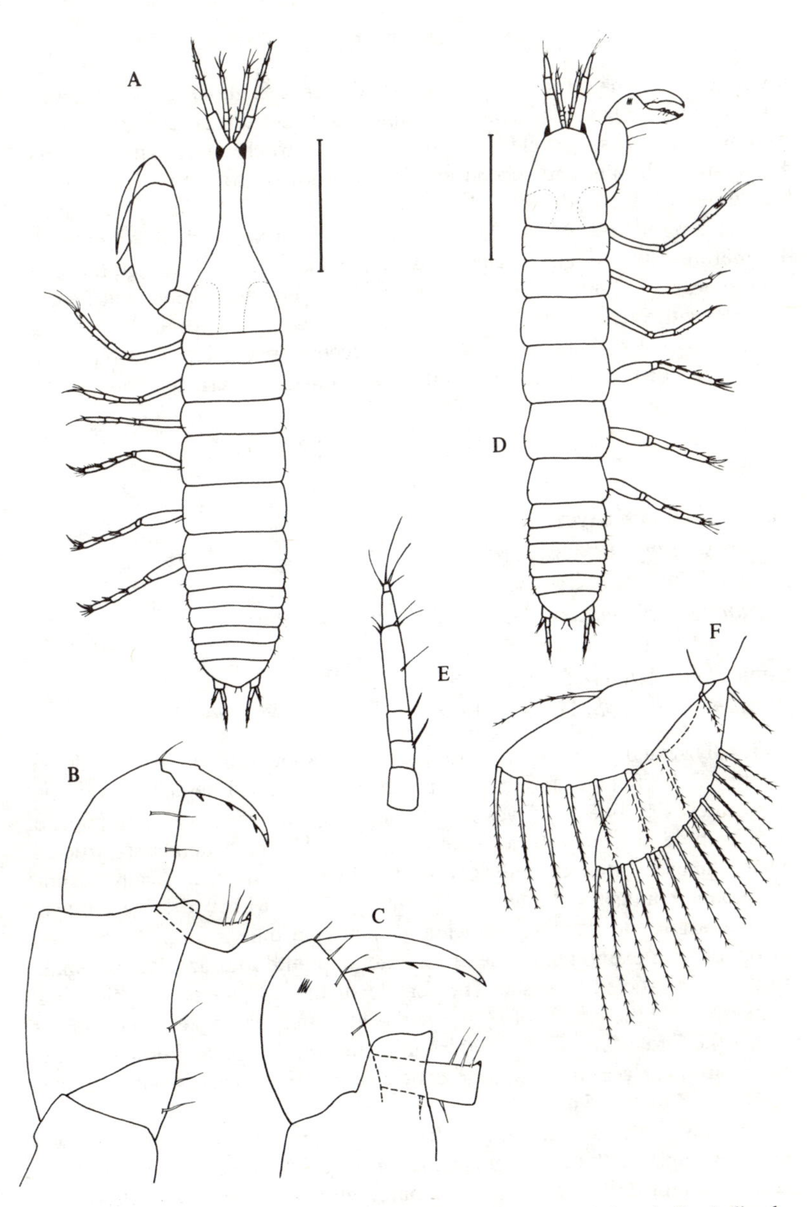

Fig. 13. *Heterotanais oerstedi*. A, male, dorsal view (scale bar 0.5 mm); B, cheliped, primary copulatory male; C, cheliped, secondary copulatory male, type B; D, female, dorsal view (scale bar 0.5 mm); E, antenna; F, pleopod, female. (B and C after Marchand, 1977).

Genus *LEPTOCHELIA* Dana, 1849

Antennule of female three- or four-articled with a single-articled flagellum, that of male with a two-articled peduncle and a multi-articled flagellum. Antenna five- or six-articled, second and third articles each with a strong dorsal spine distally and second article also with ventral spine (Fig. 14D). Mandibles with cylindrical and fluted molar process. Endite of maxillule with 11 long spines. Maxillipeds of female well developed, those of male degenerate with a single-articled palp. Chelipeds of female normal, those of male powerfully built and longer than those of female. Propodus of female cheliped with vertical row of setae, that of male with a horizontal row. Bases of pereopods 4–6 thicker than those of preceding pairs. Uropods biramous with single- or two-articled exopodite and a four–seven articled endopodite.

Leptochelia savignyi (Krøyer, 1842)

(Fig. 14)

Tanais dubius Krøyer, 1842
Tanais savignyi Krøyer, 1842
Tanais Edwardsii Krøyer, 1842
Leptochelia Edwardsii: Bate and Westwood, 1868
Leptochelia dubia: Sars, 1882
Leptochelia Savignyi: Sars, 1882
Paratanais algicola Harger, 1878

Markedly sexually dimorphic with male being approximately two-thirds the length of the female and having a differently shaped cephalothorax and pereonites (Fig. 14A, C). Eyes prominent, particularly in male. Antennule of female three-articled with a single-articled flagellum. Antenna six-articled. Male cheliped often exhibiting variation in form which has caused some taxonomic confusion (Lang, 1973), but propodus with two large teeth on inner margin and dactylus with many spines on inner margin (Fig. 14B). Uropods with exopodite single-articled; endopodite four-articled in female, five- or six-articled in male. The female of this species is very similar in appearance to the female of *H. oerstedi.* The easiest distinguishing characteristic is the form of the antenna which in *L. savignyi* has an additional spine on the ventral surface of the second article (Fig. 14D). Body length of female 2–3 mm, of male 1.5 mm.

Records of this species in the British Isles are limited to the south-west coast of England, the Channel Islands, west and south-west Ireland where, in some localities, it is a common inhabitant of the shallow sublittoral (J. Kitching, personal communication). It may also be found intertidally in self-constructed tubes amongst *Zostera* roots and weeds on rocks. *L. savignyi* is practically cosmopolitan, occurring in the Mediterranean, on the Dutch coast, and along Atlantic shores from Britanny to Senegal. There are also records from the east and west coasts of North America, Brazil, and from the Indo-West Pacific, South Africa, Hawaii and the Tuamotu Archipelago.

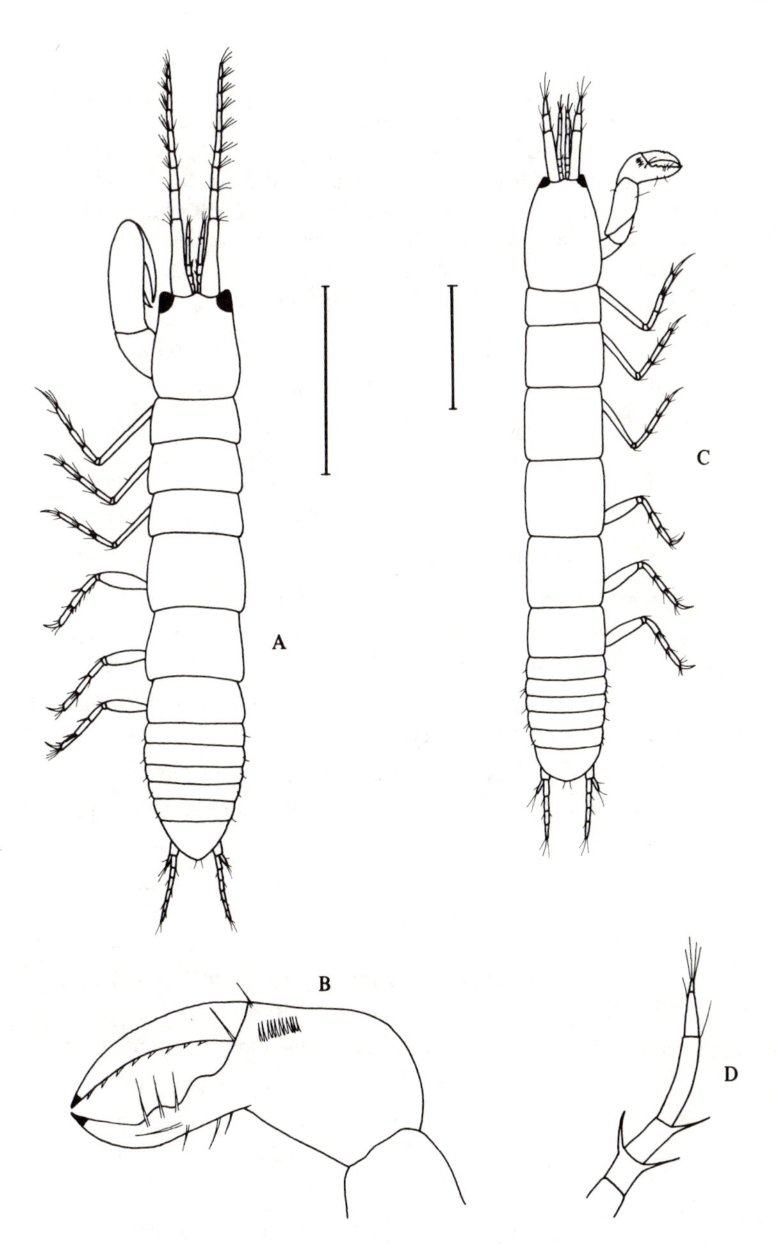

Fig. 14. *Leptochelia savignyi.* A, male, dorsal view (scale bar 0.5 mm); B, chela, male; C, female, dorsal view (scale bar 0.5 mm); D, antenna.

Family LEPTOGNATHIIDAE Sieg, 1976

Antennule of female three- or four-articled, that of male four–seven articled. Antenna five–seven articled. Molar process of mandible well developed or reduced. Maxillipedal bases exhibit varying degrees of fusion throughout the family. Labium formed from only two lobes, cleft medially. Mouthparts of male reduced except for remains of maxilliped. Chelipeds of male similar to those of female. Pereopods 1–6 with distinct dactylus and a terminal unguis. Pleopods may be present or absent and are highly variable across the genera of this family. Marsupium formed from four pairs of oostegites.

Key to British genera and species of Leptognathiidae

1. Antennule of female three-articled, of male six-articled **2**

Antennule of female four-articled, of male four- or seven-articled... **7**

2. Eyes present .... *Pseudoparatanais batei* (p. 52)

Eyes absent .... (*Typhlotanais*) **3**

3. Pereonite 1 longer than cephalothorax (Fig. 20A) .... *Typhlotanais pulcher* (p. 62)

Pereonite 1 much shorter than cephalothorax **4**

4. Anterior margin of carapace with acutely produced rostrum (Fig. 18A) .... *Typhlotanais microcheles* (p. 58)

Carapace not produced as above **5**

5. Pereopods 1–3 each with very long seta from ischium .... *Typhlotanais tenuicornis* (p. 54)

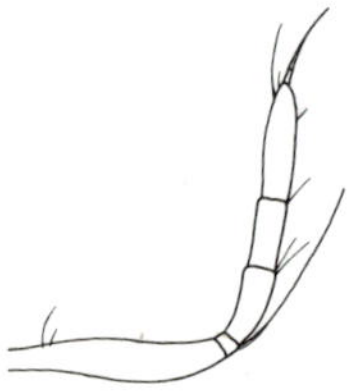

Pereopods 1–3 without very long seta from ischium **6**

6. Bases of pereopods 4–6 extremely swollen; antennules as long as carapace.................... .... *Typhlotanais aequiremis* (p. 60)

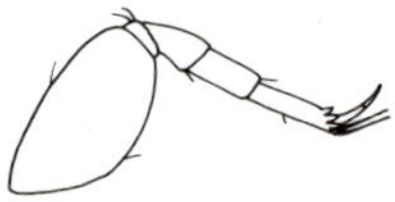

Bases of pereopods 4–6 only slightly swollen; antennules half length of carapace .... *Typhlotanais brevicornis* (p. 56)

7. Claw of cheliped propodus comprising two teeth between which dactylar unguis fits........ .... *Tanaopsis graciloides* (p. 64)

Claw of cheliped propodus simple **8**

8. Pleopods present in both sexes; male antennule seven-articled...... **9**

Pleopods absent in female, present in male; male antennule four-articled.. **12**

9. Exopodite of uropod distinctly defined from basis....................... **10**

Exopodite of uropod not distinctly defined from basis **11**

10. Exopodite of uropod two-articled, dactylus of chela serrated on proximal region of upper margin (Fig. 24H) ...... *Leptognathia gracilis* (p. 70)

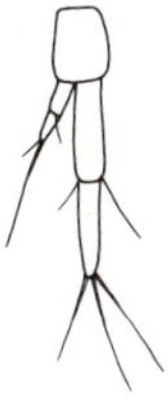

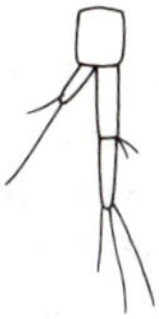

Exopodite of uropod single-articled, dactylus of chela not serrated on upper margin .... *Leptognathia breviremis* (p. 68)

11. Exopodite of uropod represented by a pointed projection indistinguishable from basis .... *Leptognathia brevimana* (p. 66)

Exopodite of uropod reduced to a small knob-like projection of the basis with two terminal setae .... *Leptognathia filiformis* (p. 76)

12. Uropod with single-articled exopodite, pleonites and pleotelson together shorter than pereonites 1 and 2 combined (Fig. 25A) .... *Leptognathia manca* (p. 72)

Uropod lacking exopodite, pleonites and pleotelson together as long as pereonites 1–4 inclusive (Fig. 26A) .... *Leptognathia paramanca* (p. 74)

Genus *PSEUDOPARATANAIS* Lang, 1973

Eyes present, well developed in male. Antennule of female three-articled, that of male six-articled. Antenna five-articled, the second article with a spiny seta ventrally, the third article with one dorsally. Chelipeds lacking pronounced sexual dimorphism. Bases of pereopods 1–3 slightly thinner than those of 4–6. Exopodites of female pleopods with superior seta (Fig. 15G). Uropod with both rami two-articled.

Pseudoparatanais batei (Sars, 1882)

(Fig. 15)

Paratanais Batei Sars, 1882
Paratanais forcipatus Bate and Westwood, 1868
Pseudoparatanais batei: Lang, 1973

Body of female linear – similar in appearance to the females of many other leptognathiid genera – body of male very much shorter and noticeably constricted in the middle of the pereon (Fig. 15A, D). Eyes much larger in male than in female. Antennules in male attaining almost one third the length of the body and inclining ventrally; proximal article of flagellum very short and, like the two following articles, with a dense bunch of aesthetascs (Fig. 15A, B). Maxilliped of female with long seta arising from just below articulation of palp (Fig. 15E). Exopodite of uropod slightly shorter and narrower than endopodite. Body length of female 1–2 mm, of male ≃ 1 mm.

This species has a vertical distribution from the littoral down to 200 m. It is often found in kept holdfasts and offshore among hydroids and in muddy sand and gravel. It is also common in washings from lower shore algae, oyster shells and water-logged wood in south-west England. Hammond (1974) recorded *P. batei* as the only common tanaid in Norfolk, and Moore (1973) lists *P. batei* as pollution intolerant, also mentioning that it is found predominantly in turbid waters. This species has been recorded from the Shetlands, the east and west coasts of Scotland, east and south-west coasts of England, western Ireland, Isle of Man and the Channel Islands. On a world-wide level records exist for the Mediterranean, the Atlantic coast of France, the Faroes, Norway and south of Iceland.

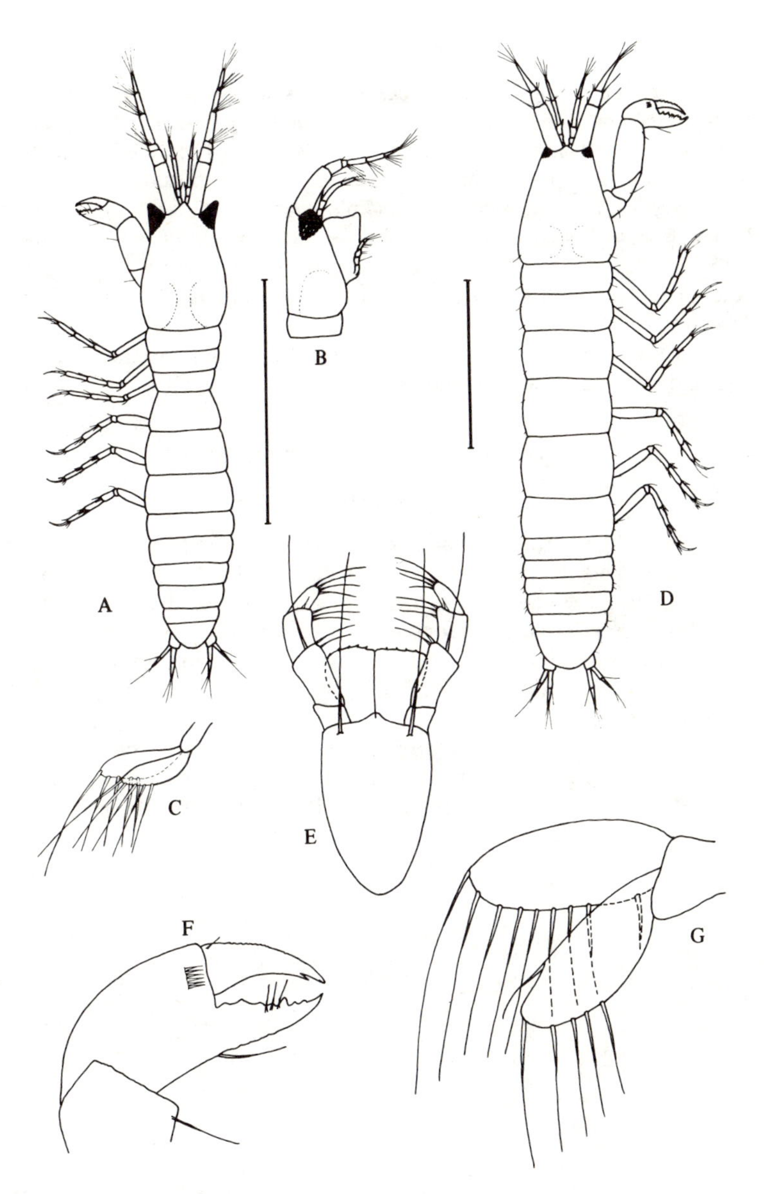

Fig. 15. *Pseudoparatanais batei.* A, male, dorsal view (scale bar 1 mm); B, lateral view of anterior end of male; C, pleopod, male; D, female, dorsal view (scale bar 1 mm); E, maxilliped, female; F, chela, female; G, pleopod, female.

Genus *TYPHLOTANAIS* Sars, 1882

Body of female slender. Ocular lobes and eyes absent. Antennule of female conical in shape, three-articled with the second article small. Antennule of male with single-articled peduncle and five-articled flagellum. Antenna six-articled. Mouthparts of male degenerate except for maxillipeds. Those of female well-developed with left mandible bearing a strong lacinia mobilis. Molar process of mandibles cylindrical, slightly dilated distally and tipped with denticles. Pereopods 1 generally longer than successive pairs. Bases of pereopods 4–6 often swollen. Pleopods better developed in male than in female. Uropods short and biramous.

This genus comprises chiefly deep-water tube-dwelling species. Males are rare and even unknown for some species.

Typhlotanais tenuicornis Sars, 1882

(Fig. 16)

Pereonite 1 slightly shorter than pereonite 6 (Fig. 16A). Antennules as long as carapace and unusually narrow with the basal article longer than the remaining two together. Third article of antenna armed with two curved teeth (Fig. 16C). Ischia of pereopods 1–3 each bear a long seta which projects at least to the distal end of the carpus (Fig. 16D). Uropods short, both rami one-articled with endopodite almost twice length of exopodite (Fig. 16B). Body length 1.5 mm.

In British waters *T. tenuicornis* has been recorded from Buckie, eastern Scotland. It is found along the Norwegian coast and has also been dredged off western Ireland at 576 m. It occurs at depths from 90–1100 m.

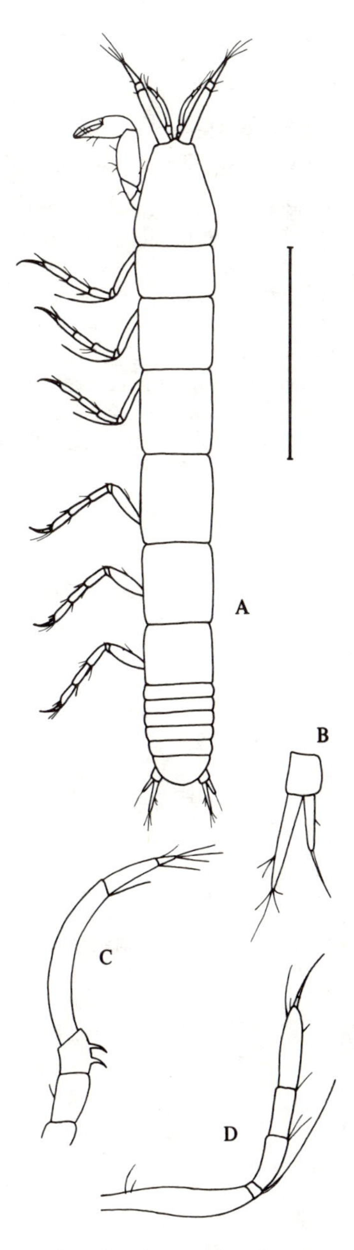

Fig. 16. *Typhlotanais tenuicornis*. A, female, dorsal view (scale bar 0.5 mm); B, uropod, female; C, antenna, female; D, pereopod 1, female.

Typhlotanais brevicornis (Lilljeborg, 1864)

(Fig. 17)

Tanais brevicornis Lilljeborg, 1864
Typhlotanais brevicornis: Sars, 1896

Pereonite 1 longer than pereonite 6 (Fig. 17A). Antennules short, scarcely half the length of the cephalothorax; basal article much longer than other two combined (Fig. 17A). Pereopods 4–6 short with bases slightly swollen. Uropods short, endopodite two-articled, exopodite very small and single-articled – generally directed outwards perpendicular to inner ramus (Fig. 17B). Body length 1.3 mm.

This species has been recorded from the east and west coasts of Scotland. It is a common inhabitant of the Norwegian coast, where it has been found on flat muddy substrates, and has a depth range from 90–870 m.

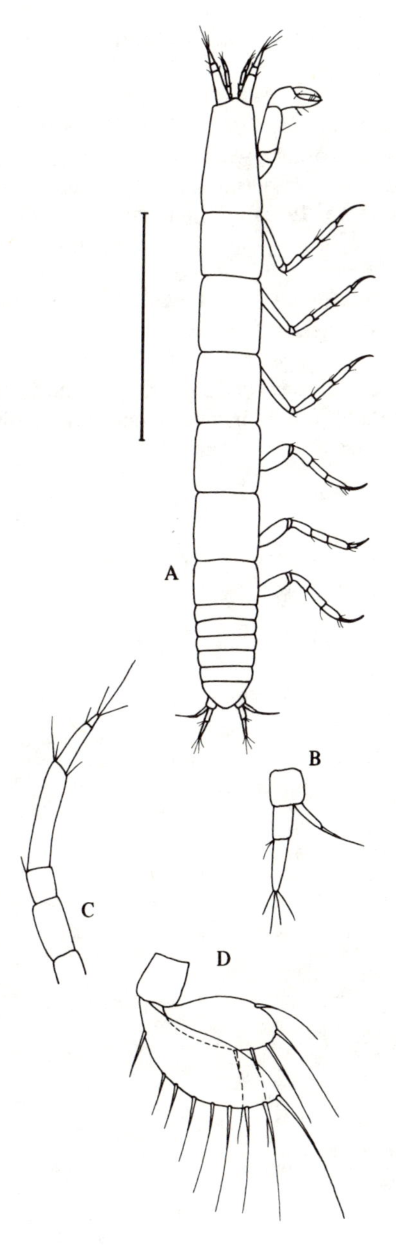

Fig. 17. *Typhlotanais brevicornis*. A, female, dorsal view (scale bar 0.5 mm); B, uropod, female; C, antenna, female; D, pleopod, female.

Typhlotanais microcheles Sars, 1882

(Fig. 18)

Frontal margin of carapace produced to an acute rostrum (Fig. 18A). Pereonite 1 shorter than others, widening anteriorly, and with prominent but thinly chitinised hook ventrally (in Norwegian specimens hooks are present on all pereonites (Greve, 1972)). Antennule slightly over half the length of the cephalothorax. Chelipeds poorly developed, propodus small and narrow (Fig. 18A). Merus, carpus and propodus of pereopods 3–6 each armed with two serrate denticles (Fig. 18D). Exopodite of uropod single-articled – almost as long as first article of the two-articled endopodite (Fig. 18B). Body length 1.5 mm.

T. microcheles has been recorded from the Isle of Man and the west coasts of Ireland and Scotland in fine or muddy sand. In addition it occurs on the coast of Norway. It has a depth range of 9–210 m.

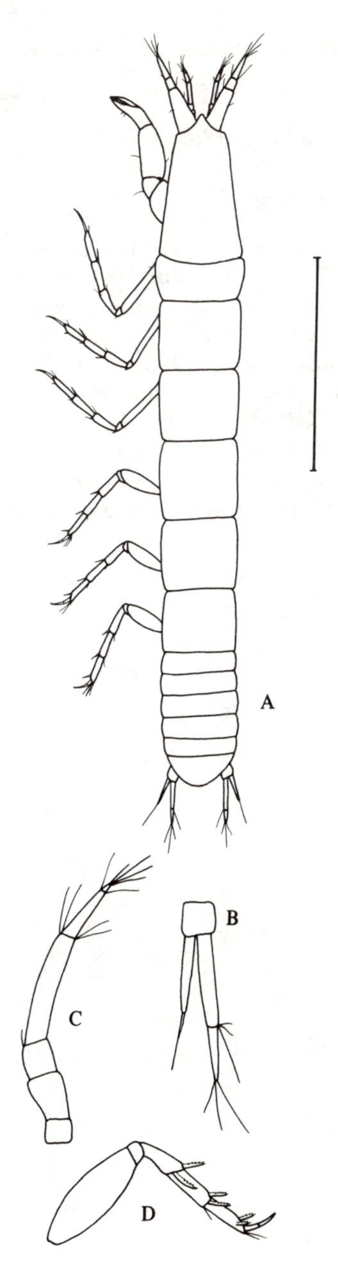

Fig. 18. *Typhlotanais microcheles*. A, female, dorsal view (scale bar 0.5 mm); B, uropod, female; C, antenna, female; D, pereopod 6, female.

Typhlotanais aequiremis (Lilljeborg, 1864)

(Fig. 19)

Tanais aequiremis Lilljeborg, 1864
Tanais depressus Sars, 1872
Typhlotanais aequiremis: Sars, 1882

Pereonite 1 shorter than others (Fig. 19A). Antennule about same length as the carapace (Fig. 19A). Chelipeds with well-developed propodus. Pereopods 4–6 with the bases unusually strongly swollen (Fig. 19F). Endopodite of uropod single-articled, exopodite two-articled and rather shorter than inner ramus (Fig. 19B). Length $\simeq$ 2.5 mm.

This species has been recorded from a number of oilfields in the northern North Sea at depths of 100–160 m, where it constructs tubes of coarse sand grains. These are the first records of *T. aequiremis* in British waters but this species is common on the Norwegian coast as far north as Tromso. It has also been recorded from Sweden and western and eastern Iceland. It has a depth range of 60–430 m.

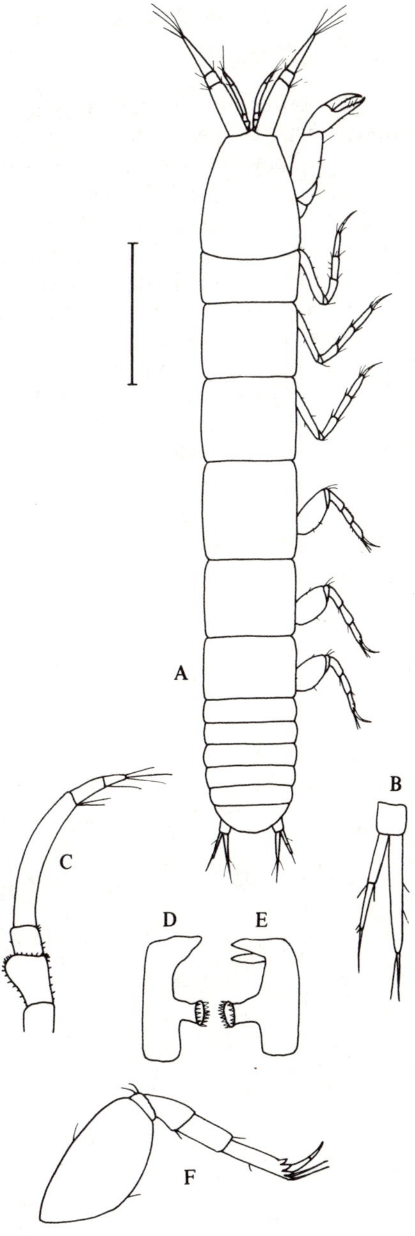

Fig. 19. *Typhlotanais aequiremis*. A, female, dorsal view (scale bar 0.5 mm); B, uropod, female; C, antenna, female; D, right mandible; E, left mandible; F, pereopod 6, female.

Typhlotanais pulcher Hansen, 1913

(Fig. 20)

Cephalothorax only slightly longer than the unusually long first pereonite which narrows posteriorly (Fig. 20A). Antennules nearly as long as cephalothorax, moderately slender. Pereonites characteristically shaped: 1–3 long and narrowing posteriorly, 4 almost parallel-sided, 5 and 6 increasing in width posteriorly (Fig. 20A). Cheliped propodus with three pointed teeth on the distal part of the incisive margin. Pereopod 1 long and slender, carpus with a distal seta almost as long as the article. Carpus of pereopods 4–6 expanded distally (Fig. 20A, B). Uropods with two-articled endopodite and single-articled exopodite which is slightly longer than the proximal article of the endopodite (Fig. 20C). Body length $\simeq$ 3 mm.

This species was based on a single specimen found by Hansen (1913) south of the Davis Straits at a depth of 3422 m. Another specimen was found recently in the Lynn of Lorn, western Scotland at a depth of 38 m.

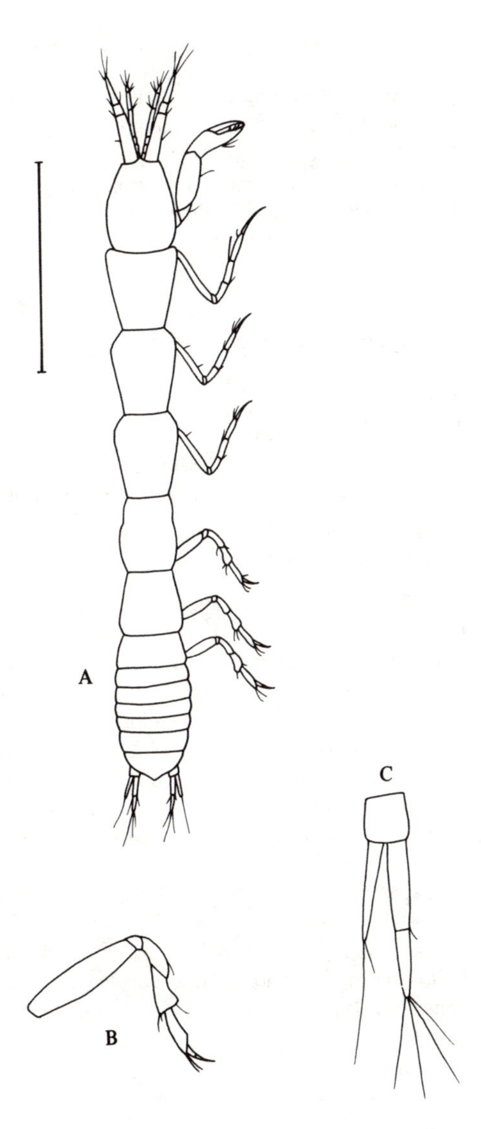

Fig. 20. *Typhlotanais pulcher*. A, female, dorsal view (scale bar 1 mm); B, pereopod 6, female; C, uropod, female.

Genus *TANAOPSIS* Sars, 1896

Body robust and eyes absent. Antennule of female four-articled. Mouthparts poorly developed. Mandibles small and rudimentary, lacking molar process (Fig. 21E, F). This genus is most easily distinguished from others of the family by the form of the maxillule which has an almost right-angled endite terminating in six spines (Fig. 21D). Chelipeds strongly built. First pair of pereopods longer than following pairs. Pleopods well developed. Uropods short and biramous.

Tanaopsis graciloides (Lilljeborg, 1864)

(Fig. 21)

Tanais graciloides Lilljeborg, 1864
Leptognathia graciloides: Norman and Stebbing, 1886
Leptognathia laticaudata Sars, 1882; Norman and Stebbing, 1886
Tanaopsis laticaudata: Sars, 1896
Tanaopsis graciloides: Lang, 1966

Body elongated, with male being much smaller than female (Fig. 21A, C). Antennule of female four-articled, that of male with two-articled peduncle and five-articled flagellum of which the three central articles bear many aesthetascs. Antenna six-articled. In the female the claw of the chelipedal propodus is two-toothed (Fig. 21H) and in the male it is usually so although it may be simple. The upper margin of the dactylus is crenulated. Pleopods well developed with exceptionally long setae in the male (Fig. 21B). Exopodite of uropod two-articled, endopodite two-articled in female and three-articled in male. The demarcation of the uropod articles is often indistinct. Body length of female ≏ 3 mm, of male ≏ 1.3 mm.

This is a common species from 5–240 m offshore where it may be found in fine and coarse sands, and sometimes in sandy mud. It has been recorded from east and west Scotland, east and south-west England, western Ireland, Isle of Man and the Shetland Islands. On a worldwide level this species is found off the coasts of Italy, northern France, Denmark, Norway and Sweden.

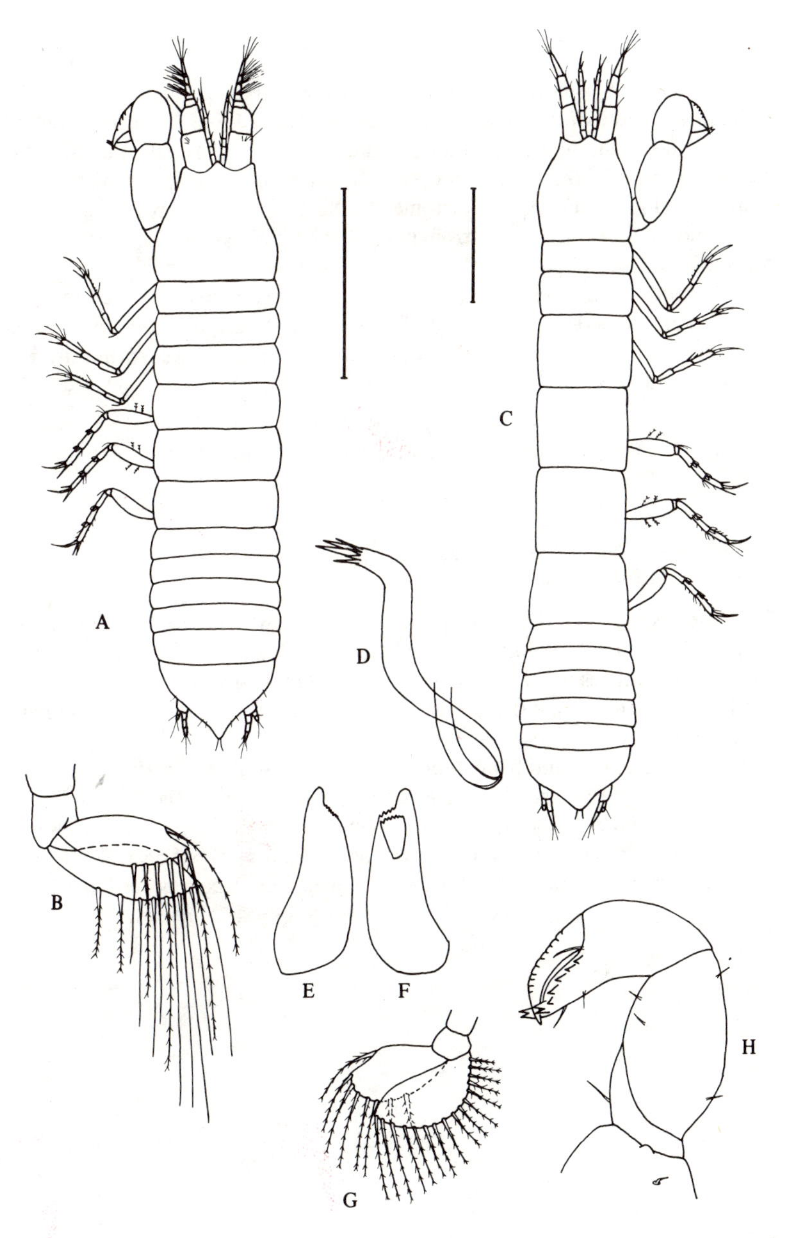

Fig. 21. *Tanaopsis graciloides*. A, male, dorsal view (scale bar 0.5 mm); B, pleopod; C, female, dorsal view (scale bar 0.5 mm); D, maxillule, female; E, right mandible; F, left mandible; G, pleopod, female; H, cheliped, female.

Genus *LEPTOGNATHIA* Sars, 1882

Body of female usually narrow and elongated, that of male shorter and wider. Antennule of female four-articled, that of male four- or seven-articled with the flagellum bearing aesthetascs and often ventrally deflected. Mandibles small and weak, molar process thin, tapering, and tipped with small denticles (Fig. 22D, E). Epistome forming a round projecting lobe. Pereopods 4–6 usually more strongly built than pereopods 1–3. Pleopods of female small, sometimes absent, those of male well developed.

This genus comprises a large number of species many of which inhabit deep water. Some species are tube dwelling. Males are not common, being unknown in some species, and keys are of necessity based primarily on female characters.

Leptognathia brevimana (Lilljeborg, 1864)

(Figs. 2C; 22)

Tanais brevimanus Lilljeborg, 1864
Leptognathia brevimana: Sars, 1882

This species is easily distinguished by the unusual structure of its uropods (Fig. 22B). The endopodite is well developed and two-articled but the exopodite is represented by a pointed projection which is not defined from the basis. Extensive material examined by the authors contained only female specimens. Body length 2–3 mm.

L. brevimana is found from depths of 17–140 m in British waters in muddy sand. It has been recorded from a number of locations on the east and west coasts of Scotland, where it is often common in sea lochs. In addition, it has been found off the coasts of north-east England, western Ireland, the Shetland Islands and in the northern North Sea. Records also exist for Norway, Sweden, Denmark and Italy down to depths of 450 m.

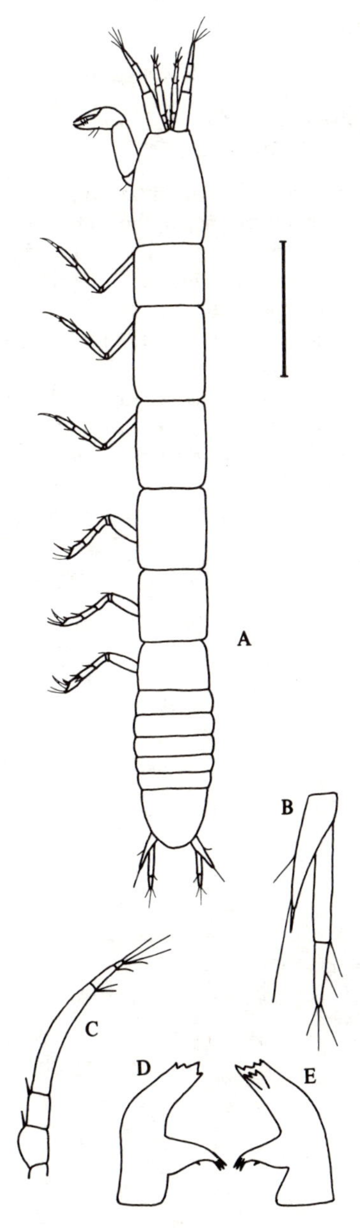

Fig. 22. *Leptognathia brevimana*. A, female, dorsal view (scale bar 0.5 mm); B, uropod, female; C, antenna, female; D, right mandible; E, left mandible.

Leptognathia breviremis (Lilljeborg, 1864)

(Fig. 23)

Tanais breviremis Lilljeborg, 1864
Leptognathia breviremis: Sars, 1882

Body short and thick in comparison with many other species of *Leptognathia* (Fig. 23A). Male similar in appearance to the male of *L. gracilis* but broader. Pleon of normal length in female, exceptionally long in male, each pleonite with a conspicuous ventral process (Fig. 23C, E). Female pleotelson terminally rounded, that of male pointed (Fig. 23C, E). Uropods short, endopodite two-articled in female (Fig. 23B), three-articled in male. Body length of female 1.3 mm, of male 1 mm.

L. breviremis has been found in muddy sand with gravel and shell off the Isle of Man. Records also exist from the east and west coasts of Scotland, Northumberland and Plymouth. It has a wide geographical distribution being found on the Swedish and Norwegian coasts, the Davis Straits (Greenland), Jan Mayen, and Iceland south to Ireland. Its depth range is from 24–3366 m.

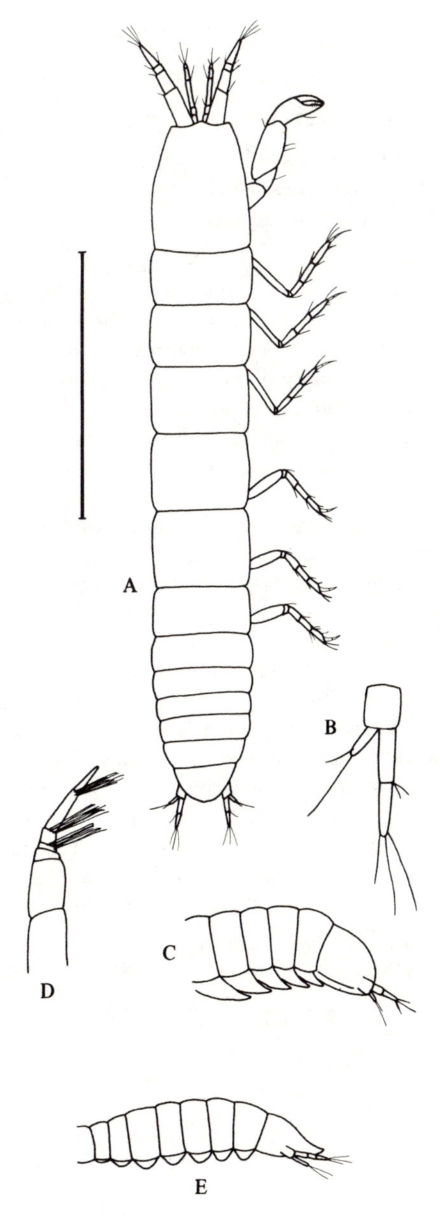

Fig. 23. *Leptognathia breviremis*. A, female, dorsal view (scale bar 0.5 mm); B, uropod, female; C, lateral view of pleon, female; D, antennule, male; E, lateral view of pleon, male.

Leptognathia gracilis (Krøyer, 1842)

(Figs. 2I, J, L, N; 3B, D, I, J; 24)

Tanais gracilis Krøyer, 1842; Lilljeborg, 1864
Leptognathia gracilis: Sars, 1882
Leptognathia longiremis: Sars, 1882
non Leptognathia longiremis Lilljeborg, 1864

Body of female elongated, that of male considerably shorter and slightly constricted in the middle (Fig. 24A, E). Both sexes with a prominent ventral tubercle on each pleonite. Propodus of female cheliped serrated or tuberculated on upper margin at insertion of dactylus. Inner margin of propodus distally with three or four sharp or round cusped teeth. Dactylus of cheliped serrated or tuberculated on upper margin, proximally at least (Fig. 24H). Uropods biramous. Endopodite long and two-articled, exopodite two-articled, shorter than first article of endopodite (Fig. 24F). Body length of female 2–2.5 mm, of male 1.2 mm.

This species is found predominantly in muddy sand, sometimes with small gravel or shell inclusions, at depths from 6–900 m. It has been recorded from the northern North Sea, the east and west coasts of both Scotland and northern England, the Isle of Man and western Ireland. *L. gracilis* has a wide, nearly circumpolar distribution, having been recorded from Alaska via Greenland to Novaya Zemlya. Its southern limit is approximately 52°N.

L. gracilis has frequently been confused with the similar *L. longiremis* Lilljeborg. *L. gracilis* has been shown by Lang (1957) to exhibit considerable phenotypic variation in such characters as the size of the ventral tubercles on the pleonites, the form of the telson, and the degree of tuberculation of the chelipedal propodus and dactylus. Nevertheless *L. longiremis* differs from *L. gracilis* in that the upper margin of the dactylus and propodus of the cheliped are smooth. Also the propodus of pereopod 1 in *L. longiremis* is provided with spines along its posterior margin.

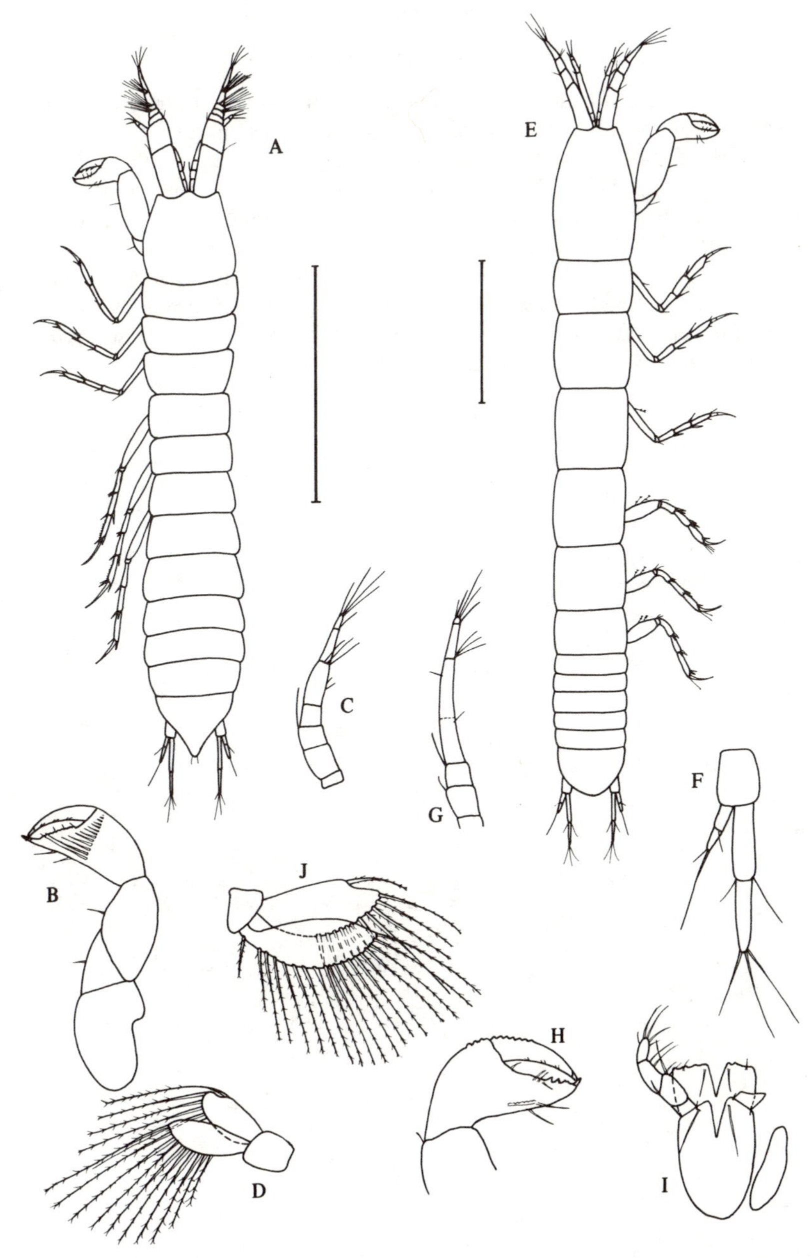

Fig. 24. *Leptognathia gracilis*. A, male, dorsal view (scale bar 0.5 mm); B, cheliped, male; C, antenna, male; D, pleopod, male; E, female, dorsal view (scale bar 0.5 mm); F, uropod, female; G, antenna, female; H, chela, female; I, maxillipeds and left maxilla, female; J, pleopod, female.

Leptognathia manca Sars, 1882

(Figs, 2D; 25)

This species is distinguished chiefly by the total absence of pleopods in the female, and by the structure of the uropods which have a two-articled endopodite and a one-articled exopodite (Fig. 25B). Body length of female 1–1.5 mm (Fig. 25A). The male of *L. manca* differs in the form of the antennules, which are thicker and longer than in the female; the pleonites are rather longer than those of the female and bear pleopods.

In the British Isles this species has been found off the Isle of Man, 6–11 km north-west of Bradda Head, where it is fairly common in muddy sand at 45–65 m depth. It has also been recorded from Norway, the Davis Straits and south of Iceland at depths of 162–1048 m.

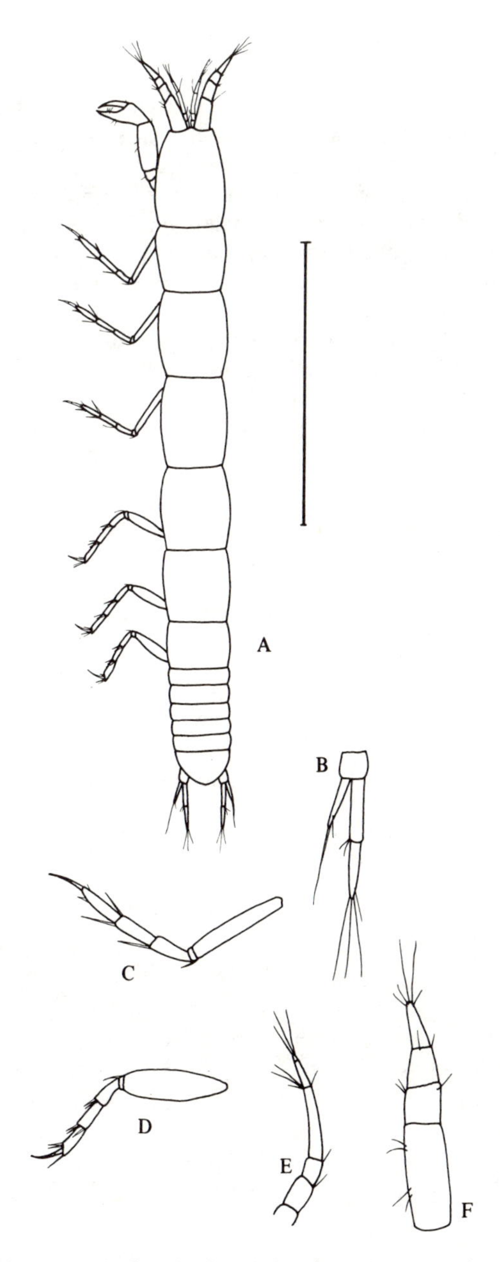

Fig. 25. *Leptognathia manca*. A, female, dorsal view (scale bar 0.5 mm); B, uropod, female; C, pereopod 1, female; D, pereopod 6, female; E, antenna, female; F, antennule, female.

Leptognathia paramanca Lang, 1958

(Fig. 26)

Body long and thin. Pleon with unusually long pleonites, pleotelson about as long as the two preceding pleonites together (Fig. 26A). Propodus and dactylus of cheliped with varying degrees of crenulation on upper margin (Fig. 26D). Pleopods absent in female as in *L. manca*. Uropods uniramous and completely lacking exopodite, endopodite two-articled (Fig. 26B). Body length of female 1.8 mm. The male of *L. paramanca* differs from the female in its possession of pleopods, the absence of mandibles, maxillules and maxillae, and by the length of the first antennular article.

This species has been recorded from off the Isle of Man at a depth of 30 m in muddy sand with gravel and shell (Bruce *et al.*, 1963). It has also been found on the north coast of France.

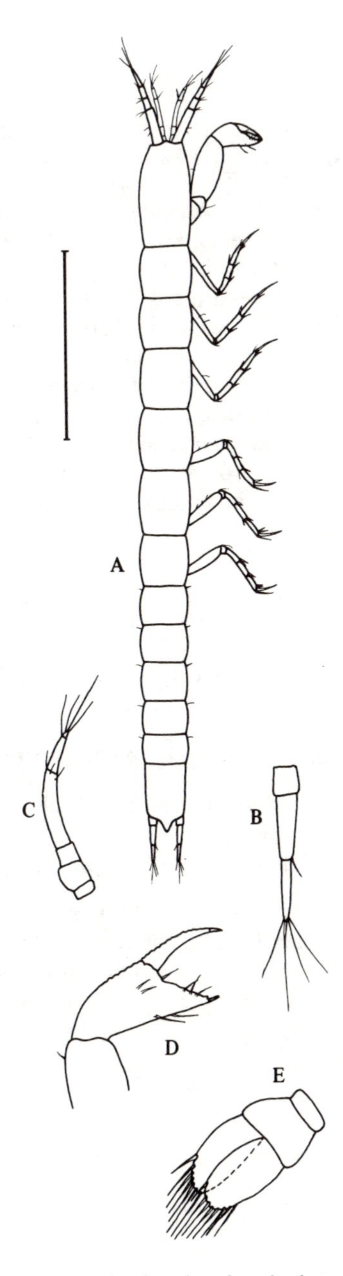

Fig. 26. *Leptognathia paramanca.* A, female, dorsal view (scale bar 0.5 mm); B, uropod, female; C, antenna, female; D, chela, female; E, pleopod, male.

Leptognathia filiformis (Lilljeborg, 1864)

(Fig. 27)

Tanais filiformis Lilljeborg, 1864
Leptognathia filiformis: Sars, 1882

Body very slender (Fig. 27A). Antennule with basal article shorter than the remaining three together. Pereopods small with the bases of pereopods 4–6 swollen. Uropods almost half the combined length of pleon and pleotelson; endopodite slender and two-articled but exopodite represented only by a very small projection of the basis (Fig. 27B). Body length 1–2 mm.

There is only one record of *L. filiformis* for the British Isles so far. McIntyre (1961) found it at a depth of 140 m at the Fladen Station, 160 km north-east of Aberdeen. There is also an unconfirmed record from a depth of 45 m off St Mary's Island, Northumberland (R. Bamber, personal communication). Records exist from the south and west coasts of Norway, where it has been found on flat muddy substrates.

SPECIES INDETERMINATA

Leptognathia rigida (Bate and Westwood, 1868)

Paratanais rigidus Bate and Westwood, 1868
Leptognathia rigida: Sars, 1882
Leptognathia rigida: Nierstrasz and Schuurmans Stekhoven, 1930

L. rigida has been listed by several authors as belonging to the British fauna. The present authors have been unable to trace any specimens of this species and on the basis of Bate and Westwood's figure alone we feel that it would be premature to place it in a definite systematic position.

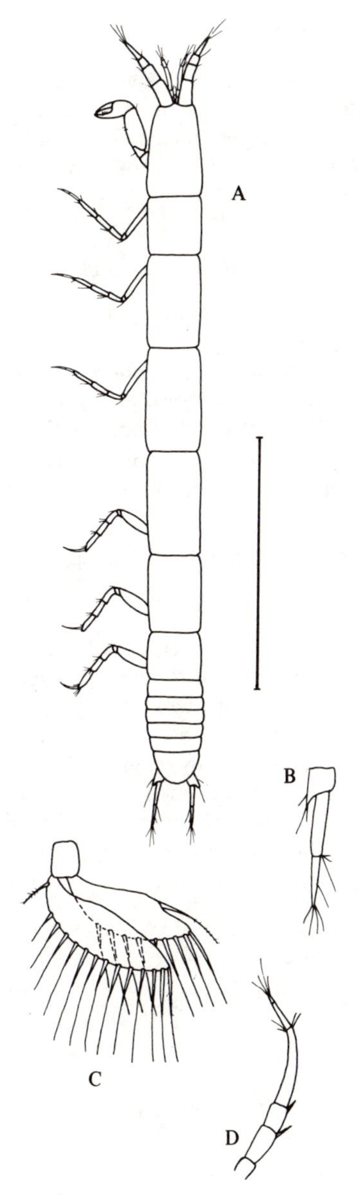

Fig. 27. *Leptognathia filiformis*. A, female, dorsal view (scale bar 1 mm); B, uropod, female; C, pleopod, female; D, antenna, female.

Family AGATHOTANAIDAE Lang, 1971

Pleon with five pleonites which are narrower than, or as broad as, the last pereonite. Ocular lobes and eyes absent. Antennule three- or four-articled. Antenna well developed or degenerate. Labium with a pennate projection from each antero-lateral corner. Chelipeds totally lacking coxa. Pleopods absent in female, present in male. Uropods markedly degenerate.

Genus *AGATHOTANAIS* Hansen, 1913

Antennule three-articled. Antenna rudimentary and single-articled (Fig. 28B), about three times as long in male as in female. Mandibles with both molar process and lacinia mobilis very reduced or absent (Fig. 28D). Bases and endites of maxillipeds fused to form a round plate with a median split anteriorly (Fig. 28C). Male pleopods tipped with a few short setae and coalesced to a pyramid-like form (Fig. 28E). Uropods reduced to a single article, not visible from the dorsal surface (Fig. 28A, E).

Agathotanais ingolfi Hansen, 1913

(Fig. 28)

Exoskeleton very brittle and completely pitted with minute round depressions. Each pereonite bears a pair of rounded protuberances at the point of insertion of the pereopods, thus giving the body a very characteristic shape (Fig. 28A). Chelipeds slender. Pereopods 3–6 each with their basis as long as the remainder of the pereopod. Body length ≏ 3 mm.

This species is newly recorded from British waters one specimen having been dredged off the Northumberland coast. Other records exist from the Davis Straits, Denmark Straits and south of Iceland at depths ranging from 1400–2158 m. This species is fairly common in grab samples taken from the Rockall area.

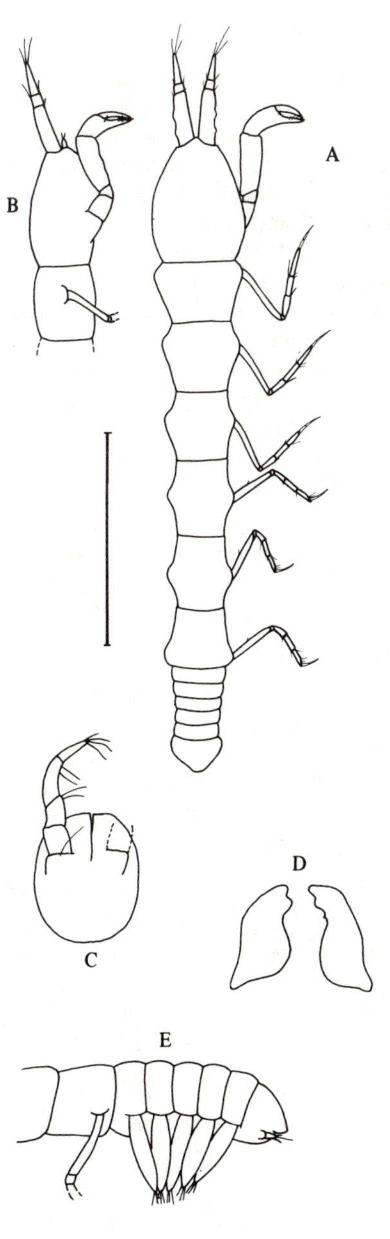

Fig. 28. *Agathotanais ingolfi.* A, female, dorsal view (scale bar 1 mm); B, lateral view of anterior of female; C, maxillipeds, female; D, mandibles, female; E, lateral view of posterior of male.

Family PSEUDOTANAIDAE Sieg, 1976

This family includes many of the smaller members of the Tanaidacea.

Antennule of female three- or four-articled, that of male with three-articled peduncle and four-articled flagellum which bears many aesthetascs (Fig. 29C). Antenna six-articled. Bases of maxillipeds fused. Lacinia mobilis of left mandible usually strong and well developed but sometimes fused to the body of the mandible. Dactylus and unguis of pereopods 4–6 may or may not be fused to form a claw. Pleopods present, reduced or absent. Uropods biramous. Marsupium formed from one pair of united oostegites.

This family contains the two subfamilies Cryptocopinae and Pseudotanainae, separable by the number of articles of the female antennule. The subfamily Pseudotanainae contains only the genus *Pseudotanais* Sars which has recently been divided into two subgenera on the basis of the form of the spine on the carpus of pereopods 2–6 (Sieg, 1977). Of the two subgenera, *Akanthinotanais* Sieg (with a normally formed spine) and *Pseudotanais s. str.*, only the latter is represented in British waters.

Key to British species of Pseudotanais

1. Pleopods absent .... *Pseudotanais forcipatus* female (p. 82)

Pleopods present **2**

2. Antennule three-articled (Fig. 30A) ... *Pseudotanais jonesi* female (p. 84)

Antennule seven-articled (Fig. 29C) **3**

3. Uropod with two-articled endopodite, one-articled exopodite (Fig. 29B) .... *Pseudotanais forcipatus* male (p. 82)

Uropod with three-articled endopodite, two-articled exopodite .... *Pseudotanais jonesi* male (p. 84)

Genus *PSEUDOTANAIS* Sars, 1882

Antennule of female three-articled. Maxillipedal bases and endites fused (Fig. 30E). Endite of maxillule with nine terminal spines. Lacinia mobilis of left mandible well developed, that of right mandible fused to body of mandible (Figs. 29H, I; 30F, G.). Mouthparts of male reduced. Dactylus and unguis of pereopods 4–6 fused to form a claw. Pleopods well developed or absent – when present endopodite lacks setae on inner margin (Figs. 29D; 30D). The subgenus *Pseudotanais* has a leaf-shaped spine on the carpus of pereopods 2–6 (Fig. 29J).

Pseudotanais (Pseudotanais) forcipatus (Lilljeborg, 1864)

(Fig. 29)

Tanais forcipatus Lilljeborg, 1864
Pseudotanais forcipatus: Sars, 1882

Cephalothorax nearly as broad as long (Fig. 29A, F). Ocular lobes and eyes absent. Pereonites 4 and 5 of female noticeably longer than other pereonites (Fig. 29F). Pleonites and pleotelson of female short (Fig. 29F). Body of male narrower than female with pereonites 3–5 longer than other pereonites (Fig. 29A). Male pleonites and pleotelson long, together only a little shorter than pereonites 1–5 together (Fig. 29A). Mandibles well developed with molar process spiniform. Chelipeds well developed in both sexes, large and robust in male (Fig. 29E, K). Pleopods absent in female, present in male. Uropods biramous. In females the exopodite and endopodite of the uropod are without distinct articles, and the exopodite is only slightly shorter than the endopodite (Fig. 29G). In the male the endopodite of the uropod is two-articled and twice the length of the single-articled exopodite (Fig. 29B). Body length of female 1.2 mm, of male 1 mm.

This species has been recorded from the east and west coasts of Scotland associated with muddy sand and mud, and from the northern North Sea in fine sand. It has a wide boreal and arctic distribution which includes the east and west coasts of Greenland, northern Iceland and the whole of the Norwegian coast. Its vertical distribution ranges from 15–349 m.

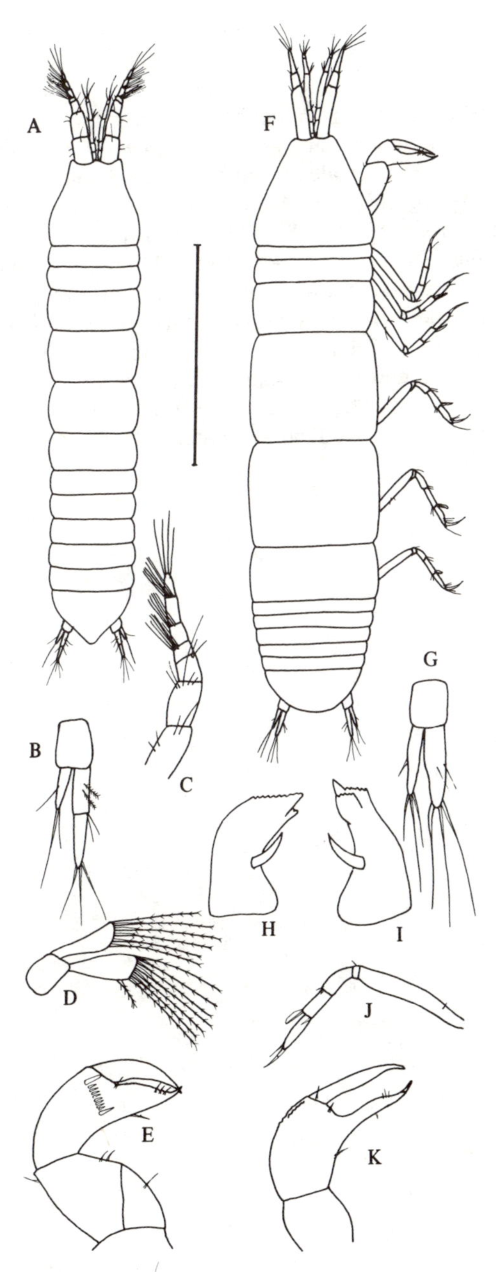

Fig. 29. *Pseudotanais* (*Pseudotanais*) *forcipatus*. A, male, dorsal view (scale bar 0.5 mm); B, uropod, male; C, antennule, male; D, pleopod, male; E, cheliped, male; F, female, dorsal view (scale bar 0.5 mm); G, uropod, female; H, right mandible, female; I, left mandible, female; J, pereopod 2, female; K, chela, female.

Pseudotanais (Pseudotanais) jonesi Sieg, 1977

(Fig. 30)

Female cephalothorax one and a half times as broad as long, triangular in shape (Fig. 30A); that of male similar to *P. forcipatus*. Ocular lobes and eyes absent. Pereonites 1–3 of female of similar size, nine times as broad as long. Pereonites 4 and 5 much longer than others (Fig. 30A). Pereon of male similar to that of *P. forcipatus*. Mandibles well developed with spiniform molar process (Fig. 30F, G). Pleopods present. Uropods of female with both rami two-articled, exopodite not as long as first article of endopodite (Fig. 30B); those of male with three-articled endopodite and a two-articled exopodite. Body length of female 0.8–1 mm, of male 1 mm.

P. jonesi has been found at one location off the Isle of Man on muddy sand and mud at depths of 32–90 m, and also in Loch Creran and the Lynn of Lorn on the west coast of Scotland associated with muddy sand and clay at depths of 20–38 m.

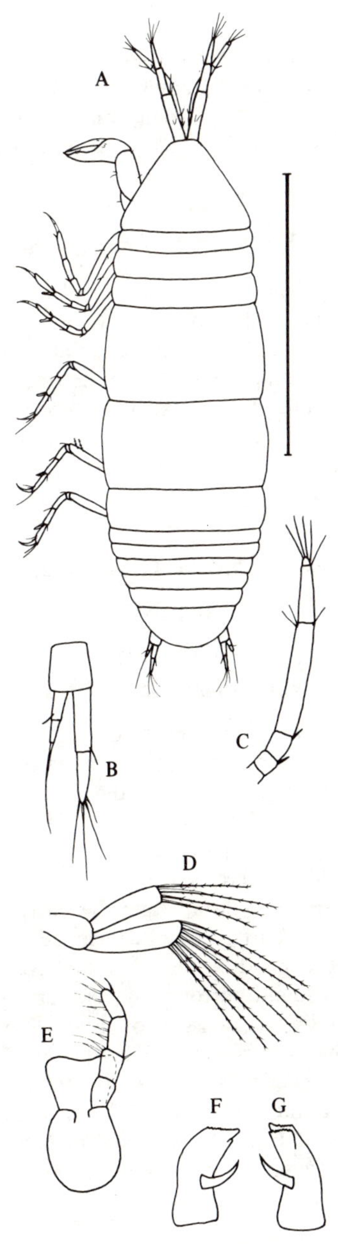

Fig. 30. *Pseudotanais* (*Pseudotanais*) *jonesi*. A, female, dorsal view (scale bar 0.5 mm); B, uropod, female; C, antenna, female; D, pleopod, female; E, maxillipeds, female; F, right mandible, female; G, left mandible, female.

Family NOTOTANAIDAE Sieg, 1976

Antennule of female three- or four-articled, that of male five- or six-articled. Antenna six-articled. Maxillipedal bases and endites fused. In the male all mouthparts exhibit some degree of reduction. Pereopods 4–6 with dactylus and unguis fused to form a claw. Pleopods reduced or well developed. Uropods biramous.

Genus *TANAISSUS* Norman and Scott, 1906

Body elongated and cephalothorax constricted anteriorly. Antennules in female four-articled, six-articled in male. Antenna six-articled but shorter than the basal article of the antennule. Maxillule strongly curved distally. Penultimate article of maxillipedal palp twice as long as terminal article (Fig. 31C). Chelipeds exhibiting strong sexual dimorphism. Although only one species of this genus has been described from British waters, a new species from the Shetlands has recently been found. This is currently being described by the authors.

Tanaissus lilljeborgi (Stebbing, 1891)

(Fig. 31)

Leptognathia Lilljeborgi Stebbing, 1891
Tanaissus Lilljeborgii Norman and Scott, 1906

Cephalothorax narrowing anteriorly (Fig. 31A). Pleotelson terminally pointed in male. Eyes absent. Antennule four-articled in female, with first article longer than combined length of the other articles, fourth article extremely small (Fig. 31A). Male antennule six-articled (Fig. 31F). Cheliped large – in the female the distal tip of the dactylus rests between the propodal unguis and two tooth-like tubercles as shown in Fig. 31A. In the male the propodus is greatly expanded distally and the dactylus is long (Fig. 31B). Pleopods well developed. Uropods well developed, both rami two-articled, exopodite about half the length of the endopodite. Body length 2–2.5 mm.

This species is often found burrowing in the top layers of intertidal and sublittoral sands, sometimes at densities of up to 600 m^{-2} (Withers, 1979). It has been found all the year round at Filey (S. McGrorty, personal communication) and at Minehead (Boydon *et al.*, 1977). *T. lilljeborgi* has been recorded from the east and west coasts of Scotland and England, the south and south-west coasts of England, Shetland and the Channel Islands. In addition, it has been found on the coasts of Denmark, the Netherlands, Brittany and Normandy.

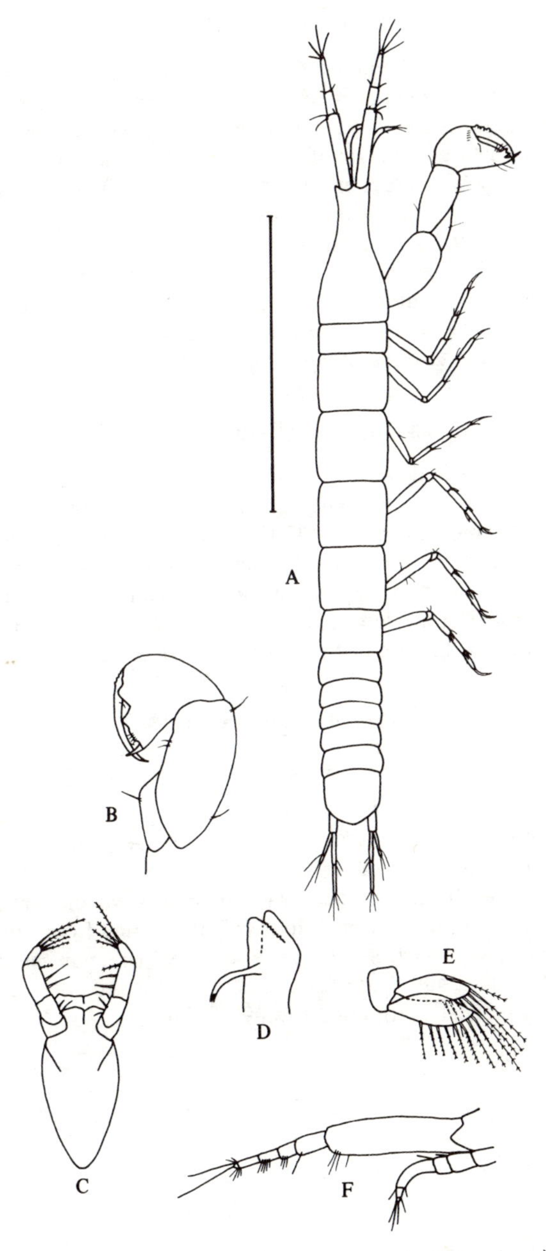

Fig. 31. *Tanaissus lilljeborgi.* A, female, dorsal view (scale bar 1 mm); B, cheliped, male; C, maxillipeds, female; D, left mandible, female; E, pleopod, female; F, lateral view of antennule and antenna, male.

Family ANARTHRURIDAE Lang, 1971

Pleon one or six-articled. Pleonites of females as broad as, or narrower than, the last pereonite. Antennule three- or four-articled, antenna six-articled. Bases of maxillipeds totally fused, endites not so. Coxa of cheliped large and articulating with proximal margin of basis (Fig. 32B). Dactylus and unguis of pereopods 4–6 may be distinct or fused to form a claw. Pleopods present or absent. Uropods short, uni- or biramous.

Following Sieg (1978b) this family is divided into two subfamilies, the Anarthrurinae and Nesotanainae, on the basis of presence or absence of eyes, the form of the antennule, the mandibular molar process and the pereopods. It is to the Anarthrurinae that the sole British representative of the Anarthruridae, *Anarthrura simplex* Sars, belongs.

Genus *ANARTHRURA* Sars, 1882

Cephalothorax short and obtusely truncated anteriorly. Ocular lobes and eyes absent. Pleon of female poorly developed, narrower than pereon and totally lacking segmentation (Fig. 32A). Pleon of male normal with five pleonites. Molar process of mandible reduced. Pereopods 4–6 more strongly built than 1–3. Pleopods absent in female, normally developed in male. Marsupium formed from four pairs of oostegites. Uropods imperfectly biramous with exopodite not distinct from basis (Fig. 32A).

Anarthrura simplex Sars, 1882

(Fig. 32)

Pleon of female approximately same length as pereonite 5 (Fig. 32A). Endopodite of uropods well defined, two-articled. Description otherwise as for genus. Body length of female ≃ 2.3 mm.

This species has been recorded twice in British waters, from 160 km north-east of Aberdeen at 140 m (McIntyre, 1961), and from 110 m in fine sand in the northern North Sea (J. Mair, personal communication). Other records exist for the south and west coasts of Norway and the west coast of France, with a vertical distribution from 90–270 m.

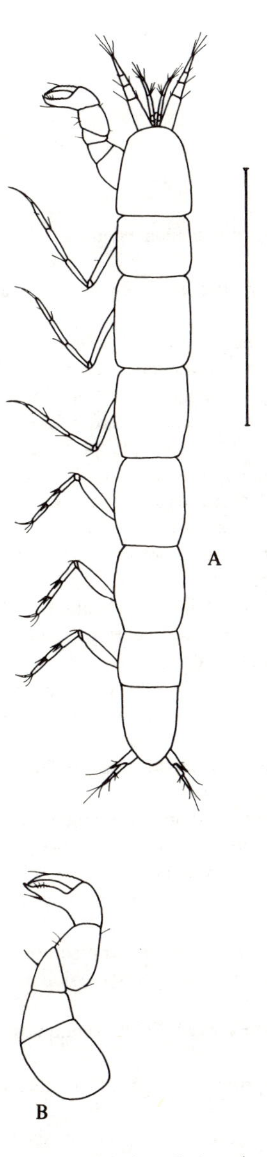

Fig. 32. *Anarthrura simplex*. A, female, dorsal view (scale bar 1 mm); B, cheliped, female.

Glossary

aesthetasc Simple, tubular, thin-walled, chemosensory seta often found on the antennular flagellum.
alimentary tract Tubular organ comprising the oesophagus, foregut, hepatopancreas and hindgut.
antenna Second pair of cephalic appendages (antenna 2).
antennal gland Excretory gland situated in the basal article of the antennal peduncle.
antennule First pair of cephalic appendages (antenna 1).
basis Article 2 of thoracic and abdominal appendages.
biramous Having two branches or rami (exopodite and endopodite).
branchial chamber A chamber limited primarily by the lateral folds of the carapace, the inner surface of which performs a respiratory function.
brood sac A pouch formed from either one or two oostegites.
carapace An extension of the exoskeleton which covers and fuses with the head and the first two thoracic somites.
carpus Article 5 of pereopod (article 4 of cheliped).
cephalon Anterior region of the body bearing the antennules, antennae, eyes, mandibles, maxillules and maxillae.
cephalothorax Cephalon and first two thoracomeres.
chelate Pincer-like condition in which dactylus closes against fixed forward projection of propodus.
cheliped Second pair of thoracic appendages; chelate in form.
chromatophore Cell containing pigment.
coxa Article 1 of thoracic and abdominal appendages.
dactylus Article 7 of pereopod (article 6 of cheliped).
endite Inwardly directed projection arising from the basis of an appendage.
endopodite (endopod) Main branch of an appendage issuing from the basis.
epignath *See* epipodite.
epimeron Lateral projection of pereonite or pleonite.
epipodite Branchial exite borne by the basal region of an appendage (especially the maxilliped).
epistome Front part of head above labrum, sometimes produced anteriorly into a spine.
exite Outwardly directed projection arising from the basal region of an appendage.
exopodite (exopod) Lateral branch of the basis.
eyes Compound structures borne on small anteriorly directed lobes of the carapace.
flagellum The often long and multi-articled region distal to the peduncle of the antennule and antenna, and the basis of the uropod.
genital cone Conical structure situated on the ventral surface of pereonite 6 on to which the male gonopore opens.
gonochoristic Pertaining to a unisexual individual; remaining as the same sex throughout the life cycle.
gonopore External opening of the oviduct or vas deferens.

head *See* cephalon.
hepatopancreas Comprising a number of caeca lying alongside the hindgut which are involved in producing digestive enzymes and in absorbing the products of digestion.
hermaphroditic Possessing functional male and female genital organs – either simultaneously or sequentially.
incisor process The fixed distal 'cutting blade' of the mandible.
ischium Article 3 of pereopod (not always present in tanaids).
labium (paragnath, lower lip) Flat, non-segmented, bilobed structure situated posterior to the oral opening.
labrum (upper lip) A bulbous structure lying next to the epistome and bordering the anterior margin of the oral opening.
lacinia mobilis The articulated 'cutting' blade of the mandible. Exhibits varying degrees of development in the Tanaidacea.
manca stage A post-embryonic stage in development; manca II leaves the marsupium.
mandibles Most anterior pair of articulated mouthparts situated on either side of oral opening.
marsupium Chamber on ventral surface of female formed by overlapping oostegites in which the eggs, embryos and mancas are brooded.
maxillae Third pair of mouthparts (maxillae 2).
maxillipeds Most posterior pair of mouthparts, derived from first pair of thoracic appendages.
maxillules Second pair of mouthparts (maxillae I).
merus Article 4 of pereopod (article 3 of cheliped).
molar process Inwardly directed process of mandible usually bearing spines or a flat grinding surface.
neutrum stage Term used by some workers to define early post-marsupial stages.
oostegites Flattened plates arising from inner proximal margin of coxa of certain pereopods.
oral opening Mouth.
ostium Opening in wall of heart through which blood is drawn from the pericardial sinus.
oviduct Tube carrying ova from ovary to the exterior.
palp Structure found on the outer margin of the mandible, maxillule, maxilla and maxilliped, sometimes reduced or absent. Articulated on mandible and maxilliped.
paragnath *See* labium.
peduncle Basal articles of antennules and antennae.
pereon Thoracic division of body, comprising six free pereonites (thoracomeres) and bearing uniramous appendages.
pereonite Somite belonging to the pereon.
pereopod Uniramous appendage of the pereon (walking leg) usually of seven articles.
pericardial sinus Space, bounded ventrally by a pericardial membrane, within which the heart lies.
pericardial membrane Muscular membrane which separates heart from rest of body cavity.
pleon Abdominal body division, usually comprising five somites and often bearing biramous pleopods.
pleonite Somite of pleon.
pleopod Biramous, often natatory, appendage of pleon.

pleotelson Pleonite 6 (sometimes 5 and 6) fused to telson.
propodus Article 6 of pereopod (article 5 of cheliped).
protandrous Used of hermaphroditism when the functional male precedes the female stage.
protogynous Used of hermaphroditism when the functional female precedes the male stage.
rostrum Anterodorsal projection of carapace.
seminal vesicle Organ which stores sperm of male.
somite Segment of the body usually differentiated into dorsal tergite and ventral sternite and with pair of appendages.
sternum Ventral surface of body.
tegumental glands Compound glands situated in pereonal cavity which secrete a silk-like substance used in construction of the animal's tube.
telson Terminal somite of the body which is always fused to pleonite 6 to form the pleotelson.
tergum Dorsal surface of the body.
thoracomere Thoracic somite, the first two of which are incorporated into the head and bear the maxillipeds and chelipeds respectively.
trituratory Chewing or grinding.
unguis Most distal article of pereopod and cheliped. In the case of a pereopod it may be fused to the dactylus to form a claw.
uniramous Having a single branch (the endopodite) or ramus.
uropods Pair of appendages borne terminally on the pleotelson, usually biramous and with multi-articled flagella.
vasa deferentia Tubes (one on either side) conveying sperm from testes to exterior.

Acknowledgements

We would like to thank the following for their generous contributions of material and, in many cases, extensive ecological data: N. S. Jones, D. McGrath, J. D. Gage, P. Kingston, R. Bamber, J. Kitching, J. P. Hartley, J. F. Tapp, R. G. Withers, A. Myers, I. Barclay, A. Eleftheriou, A. J. Newton, D. Robinson, J. Crothers, P. G. Moore, J. Hunter, A. Preston, B. Barnett and J. Mc. D. Mair. Loans of specimens were received from G. Smaldon (Royal Scottish Museum), J. M. C. Holmes (National Museum of Ireland), P. G. Oliver (National Museum of Wales), R. Oleröd (Swedish Museum of Natural History), R. J. Lincoln and J. Ellis (British Museum/Natural History). Special thanks are due to the latter for the generous provision of facilities in the Crustacea Section. We are also grateful for taxonomic discussion with R. J. Lincoln and K. Harrison, and for informative correspondence with J. Sieg, L. Greve Jensen, M. M. Parker, S. McGrorty and M. Sheader. Finally, we must thank the Scottish Marine Biological Association for use of research facilities at their Dunstaffnage Laboratory, and J. D. Gage and his assistants for their invaluable help in collecting material from Scottish sea lochs.

References

Bate, C. S. and Westwood, J. O. 1868. *A History of British Sessile-eyed Crustacea*, Vol. 2. London: J. van Voorst, 536 pp.

Boydon, C. R., Crothers, J. H., Little, C. and Mettam, C. 1977. The intertidal invertebrate fauna of the Severn Estuary. *Field Studies*, **4**(4), 477–554.

Bruce, J. R., Colman, J. S. and Jones, N. S. 1963. *Marine Fauna of the Isle of Man*. Liverpool University Press, 307 pp.

Buckle Ramirez, L. F. 1965. Untersuchungen über die Biologie von *Heterotanais oerstedi* Kröyer (Crustacea, Tanaidacea). *Z. Morph. Ökol. Tiere*, **55**, 714–82.

Charniaux-Cotton, H. 1960. Sex determination. In *The Physiology of Crustacea*, ed. T. H. Waterman, pp. 411–47. New York: Academic Press.

Claus, C. 1888. Ueber *Apseudes latreillii* Edw. und die Tanaiden, II. *Arb. zool. Inst. Univ. Wien*, **7**, 139–220.

Dana, J. D. 1852. *Crustacea*. U.S. Expl. Exped. (1838–1842). Vol. 13(2), pp. 686–1612. Philadelphia.

Dennell, R. 1937. On the feeding mechanism of *Apseudes talpa*, and the evolution of the peracaridan feeding mechanism. *Trans. R. Soc. Edinb.*, **59**, 57–8.

Gage, J. 1972. Community structures of the benthos in Scottish sea-lochs. I. Introduction and species diversity. *Mar. Biol.*, **14**(4), 281–97.

Gamble, J. C. 1970. Anaerobic survival of the crustaceans *Corophium volutator, C. arenarium* and *Tanais chevreuxi. J. mar. biol. Ass. U.K.*, **50**(3), 659–71.

Gardiner, L. F. 1973. A new species and genus of a new monokonophoran family (Crustacea: Tanaidacea), from Southeastern Florida. *J. Zool. Lond.*, **169**, 237–53.

Gardiner, L. F. 1975a. The systematics, postmarsupial development, and ecology of the deep-sea family Neotanaidae (Crustacea: Tanaidacea). *Smithson. Cont. Zool.*, **170**, 265.

Gardiner, L. F. 1975b. A fresh- and brackish-water tanaidacean, *Tanais stanfordi* Richardson, 1901, from a hypersaline lake in the Galapagos Archipelago, with a report on West Indian specimens. *Crustaceana*, **29**(2), 127–40.

Greve, L. 1972. Some new records of Tanaidacea from Norway. *Sarsia*, **48**, 33–8.

Hammond, R. 1974. The marine and brackish-water non-amphipodan peracarid crustacea of Norfolk. *Cah. Biol. mar.*, **15**(2), 197–213.

Hansen, H. J. 1895. Isopoden, Cumaceen und Stomatopoden der Plankton Expedition. II. Ordung: Tanaidacea. *Ergebn. Plankton-Exped.*, **2**(G.c.), 49–50.

Hansen, H. J. 1913. Crustacea Malacostraca, II. IV, The order Tanaidacea. In *Danish Ingolf Expedition*, Vol. 3(3), pp. 1–145. Copenhagen: H. Hagerup.

Holdich, D. M. 1968. Reproduction, growth and bionomics of *Dynamene bidentata* (Crustacea: Isopoda). *J. Zool., Lond.*, **156**, 137–53.

Holdich, D. M. and Jones, J. A. 1983. The distribution and ecology of British shallow-water tanaid crustaceans (Peracarida, Tanaidacea). *J. nat. Hist.* (in press).

Holthuis, L. H. 1956. Isopoda en Tanaidacea. *Fauna Ned.*, **16**, 1–120.

Jones, N. S. 1976. *British Cumaceans: Synopses of the British Fauna No. 8*. London: Academic Press. 66 pp.

Lang, K. 1953. *Apseudes hermaphroditicus* n. sp. A hermaphroditic tanaid from the Antarctic. *Arkiv. für Zoologie*, (2)**4,** 341–50.

Lang, K. 1956. Neotanaidae nov. fam., with some remarks on the phylogeny of the Tanaidacea. *Ark. Zool.*, (2)**9,** 469–75.

Lang, K. 1957. Tanaidacea from Canada and Alaska. *Contr. Dep. Pech., Quebec*, **52,** 1–54.

Lang, K. 1960. The genus *Oosaccus* Richardson and the brood pouch of some tanaids. *Ark. Zool.*, (2)**13**(3), 77–9.

Lang, K. 1973. Taxonomische und phylogenetische untersuchungen über die Tanaidaceen 8. Die Gattungen *Leptochelia* Dana, *Paratanais* Dana, *Heterotanais* G. O. Sars und *Nototanais* Richardson. Dazu einige Bemerkungen über die Monokonophora und ein Nachtrag. *Zool. Scr.*, **2,** 197–229.

Lauterbach, K.-E. 1970. Der Cephalothorax von *Tanais cavolinii* Milne Edwards (Crustacea – Malacostraca), ein Beitrag zur vergleichenden Anatonomie und Phylogenie der Tanaidacea. *Zool. Jb. Anat.*, **87,** 94–204.

Leach, W. E. 1814. Crustaceology. In *The Edinburgh Encyclopedia*, Vol. 7, ed. D. Brewster, pp. 383–436.

Lucas, M. 1849. Histoire Naturelle des Animaux Articulés 1. Crustacés, Arachnides, Myriapodes et Hexapodes. *Expl. sci. Algérie, Sci. Phys. Zool.*, **1,** 1–403.

McIntyre, A. D. 1961. Quantitative differences in the fauna of boreal mud associations. *J. mar. biol: Assoc. U.K.*, **41,** 599–616.

McLaughlin, P. A. 1980. *Comparative Morphology of Recent Crustacea.* San Francisco: W. H. Freeman and Co. 177pp.

Marchand, J. 1977. Observations sur la faune du canal de Caen à la Mer: Etude de la population *d'Heterotanais oerstedi* Kröyer (Crustacé, Péracaride, Tanaidacé). *Bull. Soc. Linn. Normandie*, **105,** 123–40.

Marine Biological Association. 1957. *Plymouth Marine Fauna*, 3rd edn. Plymouth.

Messing, C. G. 1979. *Pagurapseudes* (Crustacea: Tanaidacea) in south eastern Florida: functional morphology, post-marsupial development, ecology, and shell use. University of Miami, Ph.D. thesis.

Messing, C. G. 1981. Notes on recent changes in tanaidacean terminology. *Crustaceana*, **41**(1), 96–101.

Milne Edwards, H. 1840. Isopodes. *Hist. nat. Crust.*, **3,** 115–283. Pls. 31–33. Paris.

Moore, P. G. 1973. The kelp fauna of Northeast Britain. III. Multivariate classification: turbidity as an ecological factor. *J. exp. mar. Biol. Ecol.*, **13,** 127–63.

Naylor, E. 1972. *British Marine Isopods: Synopses of the British Fauna No. 3.* London: Academic Press. 90 pp.

Nierstrasz, H. F. and Schuurmans Stekhoven, J. H. 1930. Anisopoda. *Tierwelt N.-U. Östsee*, (18)**10**(3), 134–67. Leipzig.

Norman, A. M. 1899. British Isopoda Chelifera. *Ann. Mag. nat. Hist.* (7)**3,** 317–41.

Norman, A. M. and Scott, T. 1906. *The Crustacea of Devon and Cornwall.* London. 232 pp.

Norman, A. M. and Stebbing, T. R. R. 1886. On the Crustacea Isopoda of the 'Lightning', 'Porcupine', and 'Valorous' Expeditions. *Trans. zool. Soc. Lond.*, **12,** 77–141.

Salvat, B. 1967. La macrofaune carcinologique endogée des sédiments meubles intertidaux (Tanaidacés, Isopodes et Amphipodes), ethologie, bionomie et cycle biologique. *Mém. Mus. natn. Hist. nat., Paris, ser. A* (**45**), 1–275.

Sars, G. O. 1882. Revision af Gruppen: Isopoda Chelifera med. charakteristik af nye herhen hørende Arter og Slaegter. *Arch. Math. Naturv.*, **7**, 1–54. Kristiana (Oslo).

Sars, G. O. 1886. Nye bidrag til Kundskaben om Middelhavets Invertebratfauna, III: Middelhavets Saxispoder (Isopoda Chelifera). *Arch. Math. Naturv.* **11**, 263–368. Kristiana (Oslo).

Sars, G. O. 1896–1899. *An Account of the Crustacea of Norway*, 2, *Isopoda*, pp. 1–270; plates 1–100. Bergen.

Scholl, G. 1963. Embryologische Untersuchungen an Tanaidaceen (*Heterotanais oerstedi* Kröyer). *Zool. Jb.* (*Anat.*), **80**, 500–54.

Sieg, J. 1977. Taxonomische monographie der familie Pseudotanaidae (Crustacea, Tanaidacea). *Mitt. zool. Mus. Berlin*, **53**, 3–109.

Sieg, J. 1978a. Bemerkungen zur Möglichkeit der Bestimmung der Weibchen bei den Dikonophora und der Entwicklung der Tanaidaceen. *Zool. Anz.*, **200**(3/4), 233–41.

Sieg, J. 1978b. Aufteilung der Anarthrurdiae in zwei unterfamilien sowie neubeschreibung von *Tanais Willemoesi* Studer als typus-art der gattung *Langitanais* Sieg (Tanaidacea). *Crustaceana*, **35**(2), 119–33.

Sieg, J. 1980a. Sind die dikonophora eine polyphyletische gruppe? *Zool. Anz.* **205**(5–6), 401–16.

Sieg, J. 1980b. Taxonomische Monographie der Tanaidae Dana 1849 (Crustacea: Tanaidacea). *Abh. senckenb. naturforsch. Ges.*, 537, 1–267.

Sieg, J. and Winn, R. 1978. Keys to suborders and families of Tanaidacea (Crustacea). *Proc. Biol. Soc. Wash.*, **91**(4), 840–6.

Siewing, R. 1954. Morphologische Untersuchungen an Tanaidaceen und Lophogastriden. *Z. wiss. Zool.*, **157**, 333–426.

Walker, A. O. 1897. On some new species of Edriophthalama from the Irish Seas. *J. Linn. Soc.* (*Zool.*), **26**, 226–32.

Withers, R. G. 1979. Observations on the macrofauna of intertidal sands at Ryde and Bembridge, Isle of Wight. *Proc. Isle Wight Natur. Hist. Archaeol. Soc. for 1977*, **7**(2), 81–9.

Wolff, T. 1956. Crustacea Tanaidacea from depths exceeding 6000 meters. *Galathea Rep.*, **2**, 187–241.

Wolff, T. 1977. Diversity and faunal composition of the deep-sea benthos. *Nature, Lond.*, **267**, 780–5.

Index of British families, genera and species

For species and genera the correct names are in italics; synonyms are in roman. The page citations in roman are to the text; those in italics are to illustrations.